AF181414

Journey into Terahertz Radiation - Exploring the Invisible Frontier

Edited by Jinfeng Li

Published in London, United Kingdom

Journey into Terahertz Radiation - Exploring the Invisible Frontier
http://dx.doi.org/10.5772/intechopen.1003386
Edited by Jinfeng Li

Contributors
Akshita Singh, Ali-Asghar Ebrahimi, Ali Shekari, Arun Jana, Jinfeng Li, John Thiruvadigal, Jyotirmayee Dash, Nisheena Raghavan, Olena Kryvenko, Parag Seal, Seyyed Salman Seyyed-Afghahi, Sidharth Raj, Somdotta Roy Choudhury, Vijay Pillai, Vitalii Yakovliev, Vladyslav Lutsenko, Yiyang Luo, Yusuf U. Tarauni

© The Editor(s) and the Author(s) 2025

The rights of the editor(s) and the author(s) have been asserted in accordance with the Copyright, Designs and Patents Act 1988. All rights to the book as a whole are reserved by INTECHOPEN LIMITED. The book as a whole (compilation) cannot be reproduced, distributed or used for commercial or non-commercial purposes without INTECHOPEN LIMITED's written permission. Enquiries concerning the use of the book should be directed to INTECHOPEN LIMITED rights and permissions department (permissions@intechopen.com).

Violations are liable to prosecution under the governing Copyright Law.

Individual chapters of this publication are distributed under the terms of the Creative Commons Attribution 4.0 License which permits commercial use, distribution and reproduction of the individual chapters, provided the original author(s) and source publication are appropriately acknowledged. If so indicated, certain images may not be included under the Creative Commons license. In such cases users will need to obtain permission from the license holder to reproduce the material. More details and guidelines concerning content reuse and adaptation can be found at http://www.intechopen.com/copyright-policy.html.

Notice

Statements and opinions expressed in the chapters are these of the individual contributors and not necessarily those of the editors or publisher. No responsibility is accepted for the accuracy of information contained in the published chapters. The publisher assumes no responsibility for any damage or injury to persons or property arising out of the use of any materials, instructions, methods or ideas contained in the book.

First published in London, United Kingdom, 2025 by IntechOpen
IntechOpen is the global imprint of INTECHOPEN LIMITED, registered in England and Wales, registration number: 11086078, 167-169 Great Portland Street, London, W1W 5PF, United Kingdom

For EU product safety concerns: IN TECH d.o.o., Prolaz Marije Krucifikse Kozulić 3, 51000 Rijeka, Croatia, info@intechopen.com or visit our website at intechopen.com.

British Library Cataloguing-in-Publication Data
A catalogue record for this book is available from the British Library

Journey into Terahertz Radiation - Exploring the Invisible Frontier
Edited by Jinfeng Li
p. cm.
Print ISBN 978-1-83634-113-0
Online ISBN 978-1-83634-112-3
eBook (PDF) ISBN 978-1-83634-114-7

If disposing of this product, please recycle the paper responsibly.

We are IntechOpen,
the world's leading publisher of
Open Access books
Built by scientists, for scientists

7,500+
Open access books available

195,000+
International authors and editors

210M+
Downloads

156
Countries delivered to

Our authors are among the
Top 1%
most cited scientists

12.2%
Contributors from top 500 universities

CLARIVATE ANALYTICS
BOOK CITATION INDEX
INDEXED

WEB OF SCIENCE™

Selection of our books indexed in the Book Citation Index
in Web of Science™ Core Collection (BKCI)

Interested in publishing with us?
Contact book.department@intechopen.com

Numbers displayed above are based on latest data collected.
For more information visit www.intechopen.com

Meet the editor

Prof. Jinfeng Li is an Engineering Scientist and an authority on liquid crystals-based microwave, millimetre-wave, and terahertz technology. He received the Institution of Engineering and Technology (IET) Award, the Arthur Jarvis Prize, the Annals of Emerging Technologies in Computing (AETiC) Highly Cited Article Award, and three Best Paper Awards at IEEE, IOP and IET conferences, respectively. He was a Cambridge Trust Scholar, Emerging Technologist with Barclays UK, Editorial Board member of three Science Citation Index journals, and Session Chair of seven IEEE conferences. He was elected National-level Young Talent, Senior Member of the China Institute of Communications, Newton Prize reviewer for the UK National Commission for UNESCO, and Grants Reviewer for the Health and Social Care Delivery Research (HSDR) fund from the National Institute for Health Research (NIHR), UK. He teaches four undergraduate modules: (1) States of the Arts in Liquid Crystals Millimeter-wave Technology for 6G; (2) Frontiers and Progress of Electrical and Computer Engineering; (3) Frontiers of Electronic Science and Technology; and (4) States of the Arts in Piano.

Contents

Preface

The terahertz (THz) spectrum, lying between the microwave and infrared regions, has long intrigued scientists and engineers for its vast potential across diverse disciplines. Once considered an elusive frontier, recent advancements in technology and research have begun to unlock the mysteries and applications of terahertz radiation, marking the dawn of a transformative era.

This book, *Journey into Terahertz Radiation – Exploring the Invisible Frontier*, provides a comprehensive guide to understanding the burgeoning field of terahertz science. Drawing on the expertise of researchers and practitioners, it traverses the dynamic landscapes of terahertz applications, from biomedical sensing and military navigation to the study of astronomical phenomena and advanced material characterization. Each chapter is designed to illuminate this cutting-edge technology's remarkable progress and future potential.

Beginning with an introductory chapter that lays the foundation for reconfigurable terahertz systems, the book progresses through diverse topics, including the pivotal role of antennas in THz technology, the unification of astronomical observations in the infrared and submillimeter regions, and the integration of novel materials like GaN and GaAs for next-generation THz electronics. In addition, the precise measurement capabilities of terahertz systems for industrial and biomedical purposes are explored, highlighting the transformative potential of this spectral domain.

Whether you are a seasoned researcher, a curious student, or a professional seeking to harness the power of terahertz technology, this book offers rich knowledge and inspiration. We hope it will contribute to further advancements in the field and inspire new innovations to shape the future.

Jinfeng Li
Beijing Institute of Technology,
Beijing, China

Section 1

THz Reconfigurability

Chapter 1

Introductory Chapter: Establishing New Life in Terahertz Reconfigurability

Jinfeng Li

1. Introduction

A clear demarcation between 5G [1] and 6G [2] may not be defined with ease at the current landscape wherein the 5G has yet to be fully rolled out or evaluated at the millimeter-wave (mm-Wave) regime. However, the insatiable desire for bandwidth and speedier flow of data is continuously gearing up the frequency upgrade to Terahertz (THz) [3] and even visible light communication (LiFi) [4]; the co-existence of both may well pave the way to the 6G landscape to be matured in 2030 [5]. Most of the future applications (e.g., real-time holograms, immersive augmented reality (AR) and virtual reality (VR), gesture sensing, general artificial intelligence (AGI), etc.) are largely data-hungry and necessitate massively higher data transfer rates than the current 5G standard. Consequently, bandwidth expansion will remain essential. This motivates the use of sub-millimeter waves (THz) and visible light communications (highest data rates so far) in place of the current microwave and millimeter-wave technology. In addition to terrestrial services, the further application scope is distributed over divisions of meteorological service, radio telescopes, and inter-satellite communications. Arguably, the vast and promising potentials in next-generation digitalization and intelligentialization should not be excluded regarding the use of THz.

In the wake of the 26th United Nations Climate Change Conference (COP26) [6], the vision of addressing societal challenges in building a greener future with low power consumption, reduced footprint, and improved energy efficiency has fueled the move toward agility and reconfigurability in electronics, wherein reconfigurable circuits and systems have boomed in recent decades. Among a range of technologies for permittivity programmability at THz, solid-state-based reconfigurable electronics and photonics are the mainstream, as exemplified by p-i-n diodes and microelectromechanical systems (MEMS) [7]. Nevertheless, for THz, the status quo on existing (mainstream p-i-n didoes) semiconductor-based electronics and optoelectronics-based photonics (including using phase change materials [8]) is arguably being challenged by the power consumption, insertion loss, and limited resolution (due to binary switching). Furthermore, another chief drawback of the MEMS switches is the limited operational lifespan (up to 2 years) and the fabrication complexity (cost) [9].

Apart from the solid-state switching approaches, alternative reconfigurability-enabled methods leverage the use of functional materials that respond (linearly or non-linearly) to the externally applied fields (e.g., one of or a combination of

IntechOpen

electronic, magnetic, thermal, optical fields, etc.), hence self-modifying their electric or magnetic properties that perturb the wave propagation characteristics (including but not limited to wave velocity, phase, polarization, and attenuation). One such highly tunable material existing in the globe and exhibiting a low-power-consuming, low-dissipative nature is liquid crystals (LC) [10], more specifically, LC in the nematic phase (abbreviated as LC in this work). Since the triumph of LC display (LCD) more than five decades ago [11], and the inception of LC-based radiofrequency passive electronics in the 1990s [12], the applicability of this functional dielectric has extensively been studied and expanded in the spectrum.

2. Liquid crystal-based THz reconfigurability

Zooming out and looking into the full spectrum, however, there is indeed a THz gap concerning the LC applications, since LC MW devices and LC optical devices (e.g., LCD) have been developed and commercialized, respectively, with resounding success. The intersection of the Internet of Things (IoT) with LC technologies opens up exciting research avenues in both consumer and industrial applications.

In particular, LC can be used in THz chip development to enable tunable or reconfigurable electronic component integration, particularly in areas such as phase shifters, reconfigurable filters (resonance-tunable frontend module), tunable antennas (modifiable radiation patterns or resonant frequency), on-chip LC displays (lightweight, low-power wearable AR/VR), and electro-optical devices (light manipulation), as shown in **Figure 1**. With the distinct wavelength range of THz, assisted by the unique properties of LC (**Figure 2**), the correspondingly LC-combined THz chip is envisioned to rewrite the chapters of the existing technology landscape.

As highlighted in **Figure 2**, the low-power error-free beam steering becomes possible thanks to the stepless tuning nature [13] of LC, a testament to its main

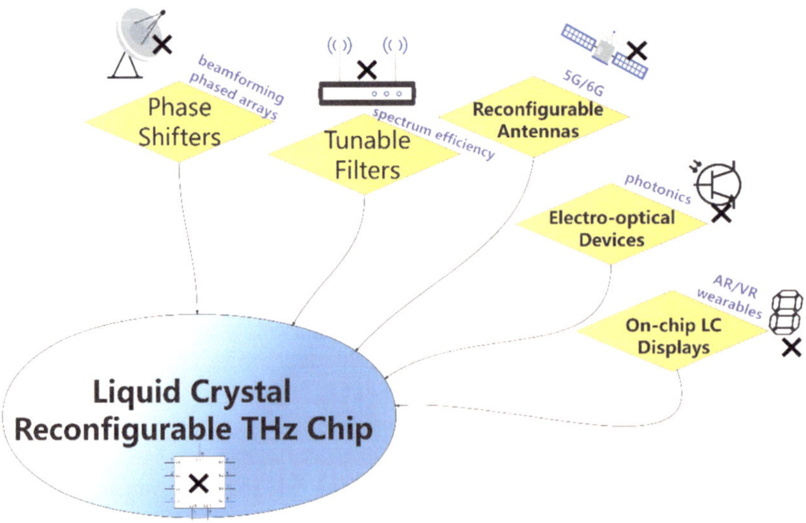

Figure 1.
Projected adoption of LC-based reconfigurable THz chips for rewriting the rules of existing technological services in commissioning.

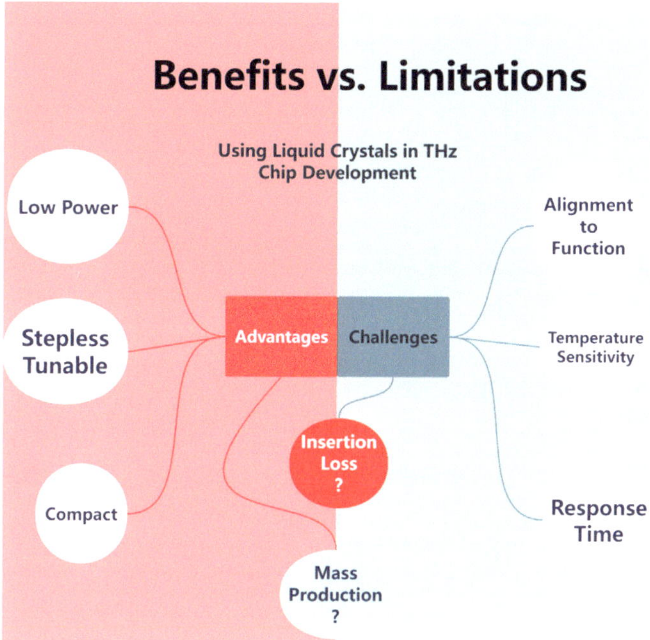

Figure 2.
Potential advantages and limitations of LC-based THz chips.

advantage. Pilot work for the LC-THz enabling technology is demonstrated by our pioneering work [14], as illustrated in **Figure 3** below, wherein coaxial transmission lines operating in TEM mode represent the most effective perturbation-sensing structure for LC-accommodation-based phase-shifting devices. Additionally, the elimination of polyimide (PI) alignment agents streamlines sample pretreatment (conventionally temperature-sensitive and complex), reducing both time and cost, which significantly enhances the feasibility of mass production (a long-standing

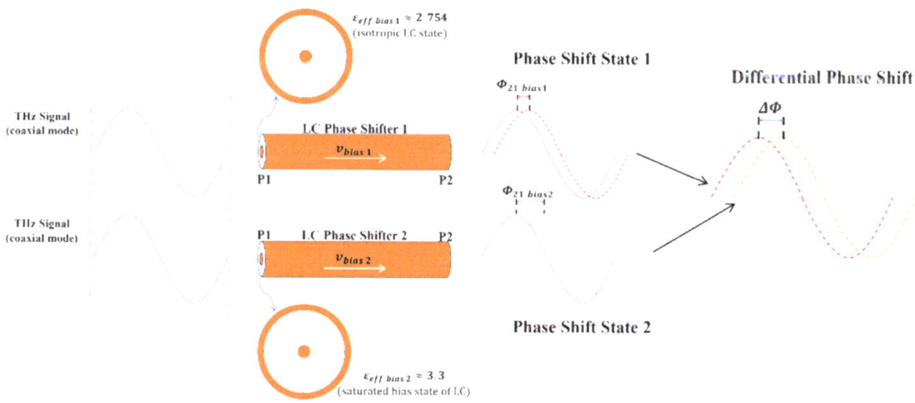

Figure 3.
Macroscopic working principle of an LC-enabled transmission line-based THz phase shifter for continuous phase modulation (exemplified by the maximum achievable tuning range between the permittivity extremes of 2.754 and 3.3 for GT3-24002 LC filled coaxially in a cavity bounded by circular conductors).

Coaxial metal loss per unit length:

$$P_{core_{s_0}} = I_{s_0}^{2}R = \left[\int_{r=\frac{D_{core}}{2}} \int_{\emptyset=0}^{\emptyset=2\pi} \sigma E_s(r,\emptyset)_0 \, d\emptyset \, dr\right]^2 \times \frac{\sqrt{\pi f \sigma \mu}}{\sigma \pi D_{core}}$$

$$P_{housing_{s_0}} = \left[\int_{r=\frac{D_{core}}{2}+T_{LC}} \int_{\emptyset=0}^{\emptyset=2\pi} \sigma E_s(r,\emptyset)_0 \, d\emptyset \, dr\right]^2 \times \frac{\sqrt{\pi f \sigma \mu}}{\sigma \pi (2T_{LC} + D_{core})}$$

LCs dielectric loss per unit length:

$$P_{LC_{s_0}} = 2\pi f \varepsilon_0 Dk_{LC} DF_{LC} \int_{r=\frac{D_{core}}{2}}^{r=\frac{D_{core}}{2}+T_{LC}} \int_{\emptyset=0}^{\emptyset=2\pi} \left|\overrightarrow{E_s(r,\emptyset)_0}\right|^2 d\emptyset \, dr$$

Analytical (phasor form)

Full-wave simulation

HFSS full – wave model:

Metal loss: $P_{core\ line} = \iint p_{core} \, dS$

$P_{housing} = \iint p_{housing} \, dS$

LCs loss: $P_{LC} = \iiint p_{LC} \, dV$

Figure 4.
Mathematically specified loss components in an LC coaxial THz transmission line.

challenge) for commercial applications. Furthermore, the PI-free coaxial structure reduces the additional insertion loss induced due to the conventional PI alignment treatment. While the dominant source of loss is attributed to the LC dielectrics, past studies only told part of the story. We mathematically summarize the different loss items in **Figure 4**, wherein the cylindrical coordinate and field denotation system can be referred to in our past works [14, 15].

3. Concluding remarks

In the evolution from 5G to 6G, one of the biggest challenges from a user's perspective is achieving a seamless switch from ground to satellite communications. By integrating LC materials into THz reconfigurable components, researchers and engineers can create highly adaptable and efficient devices, driving innovations in both THz and optical agile technologies that underpin this transition. Currently, THz (sub-mmW) LC devices are technically viable but not commercially practical. There is arguably a long road ahead to go to realize many LC-based THz research applications, pushing the boundaries of the possible and opening pathways to highly sophisticated novel technologies, albeit the untapped potential and rewards will be inestimable. While the substantial challenges and opportunities underpinning the LC-mmW combination have been briefed in the editorial [10] documented in 2022 and marched toward sub-mmW (i.e., THz) in **Figure 2** above, new development insights into tuning speed improvements are evidenced in the recent year, for example, by leveraging the device's aspect ratio effect [16]. Meanwhile, ongoing work has been started to enhance LC materials' competitiveness for shrinking the THz gap. Rethinking out of the box and seeing the convergence of LC THz reconfigurability and quantum systems [17] is another revelation to accomplish the internet-of-everything vision.

Acknowledgements

The editor acknowledges all the contributing authors to this book. The dissemination of this research is supported by the National Natural Science Foundation of China (NSFC Grant 62301043) and the Fundamental Research Funds for the Central Universities (Beijing Institute of Technology Research Fund Program for Young Scholars).

Author details

Jinfeng Li
Beijing Institute of Technology, Beijing, China

*Address all correspondence to: jinfengcambridge@bit.edu.cn

IntechOpen

© 2024 The Author(s). Licensee IntechOpen. This chapter is distributed under the terms of the Creative Commons Attribution License (http://creativecommons.org/licenses/by/4.0), which permits unrestricted use, distribution, and reproduction in any medium, provided the original work is properly cited.

References

[1] Kishk A, editor. The Fifth Generation (5G) of Wireless Communication. London, UK: IntechOpen; 2019. DOI: 10.5772/intechopen.71356

[2] Paulo S, Ramjee P, editors. 6G The Road to the Future Wireless Technologies 2030. 1st ed. New York: River Publishers; 2021

[3] Meziani Y, Pérez J, editors. Trends in Terahertz Technology. London, UK: IntechOpen; 2023. DOI: 10.5772/intechopen.102214

[4] Swami K, Moghe A. A review of LiFi technology. In: Proceedings of the 2020 Fifth IEEE International Conference on Recent Advances and Innovations in Engineering (ICRAIE); 01-03 December 2020; Jaipur, India. New York: IEEE; 2021. pp. 1-5. DOI: 10.1109/ICRAIE51050.2020.9358340

[5] Kim N, Kim G, Shim S, Jang S, Song J, Lee B. Key technologies for 6G-enabled smart Sustainable City. Electronics. 2024;**13**(2):268. DOI: 10.3390/electronics13020268

[6] Perissi I, Armenia S, Pasqualino R, Csala D. Special issue "from COP 26 to COP 27: Contributions of systems approaches to address the challenges ahead": An editorial commentary. Sustainability. 2023;**15**(18):13519. DOI: 10.3390/su151813519

[7] Abbas A, Hussain N, Sufian MA, Awan WA, Lee J, Kim N. An electronically reconfigurable highly selective stop-band ultra-wideband antenna applying electromagnetic bandgaps and positive-intrinsic-negative diodes. Micromachines. 2024;**15**(5):638. DOI: 10.3390/mi15050638

[8] Tomer D, Coutu RA. A phase change material for reconfigurable circuit applications. Applied Sciences. 2018;**8**(1):130. DOI: 10.3390/app8010130

[9] Kurmendra AS. MEMS switch realities: Addressing challenges and pioneering solutions. Micromachines. 2024;**15**(5):556. DOI: 10.3390/mi15050556

[10] Li J. Challenges and opportunities for nematic liquid crystals in radio frequency and beyond. Crystals. 2022;**12**(5):632. DOI: 10.3390/cryst12050632

[11] Jones J. The fiftieth anniversary of the liquid crystal display. Liquid Crystals Today. 2018;**27**(3):44-70. DOI: 10.1080/1358314X.2018.1529129

[12] Li J. From liquid crystal on silicon and liquid crystal reflectarray to reconfigurable intelligent surfaces for post-5G networks. Applied Sciences. 2023;**13**(13):7407. DOI: 10.3390/app13137407

[13] Li J. Rethinking liquid crystal tunable phase shifter design with inverted microstrip lines at 1-67 GHz by dissipative loss analysis. Electronics. 2023;**12**(2):421. DOI: 10.3390/electronics12020421

[14] Li J, Li H. Modeling 0.3 THz coaxial single-mode phase shifter designs in liquid crystals with constitutive loss quantifications. Crystals. 2024;**14**(4):364. DOI: 10.3390/cryst14040364

[15] Li J, Li H. Liquid crystal-filled 60 GHz coaxially structured phase shifter design and simulation with enhanced figure of merit by novel permittivity-dependent impedance matching. Electronics. 2024;**13**(3):626. DOI: 10.3390/electronics13030626

[16] Li J, Li H. Navigating aspect ratio effects in response time challenges of liquid crystal coaxial phase shifters for next-generation mmW communications. In: Proceedings of the 2024 11th International Conference on Wireless Networks and Mobile Communications (WINCOM); 23-25 July 2024; Leeds, United Kingdom. New York: IEEE; 2024. pp. 1-5. DOI: 10.1109/WINCOM62286.2024.10655435

[17] Rodarte AL, Cisneros F, Hein JE, Ghosh S, Hirst LS. Quantum dot/liquid crystal nanocomposites in photonic devices. Photonics. 2015;**2**(3):855-864. DOI: 10.3390/photonics2030855

Section 2

THz Electronics

Chapter 2

GaN, InAs, GaAs, Terahertz (THz) Electronics for Bio-Medical Sensing and Military Navigation Applications

Yusuf U. Tarauni and John Thiruvadigal

Abstract

Consistent use of silicon or graphene and other semiconductor-based materials for terahertz generation and detection leads to a plain effect such as the device running out of steam above the frequency of ~100 GHz due to the transport time limitations for the device to operate at high terahertz frequencies that range from 0.1 to 10 THz, which demarcates electronics from photonics and has, over a long period, been difficult to access and process. Also, photonic devices falter below a frequency of ~10 THz, which forces the thermal energy of the phonon to decrease down to room temperature. Therefore, the research society needs alternative materials using homojunctions and heterojunctions of III-V-based semiconductors like GaN and GaAs as materials for terahertz detectors. A clear response is needed at that upper part of the frequency range. A new strategy is required using a recessed-gated III-V semiconductor-based high electron mobility transistor to operate in the terahertz region.

Keywords: terahertz, GaAs, CMOS, 2DEG, HEMT, teraelectronics

1. Introduction

The terahertz (THz) wave in the frequency spectrum can be termed as the submillimeter range within the wavelength of 300 GHz to 10 THz in the electromagnetic spectrum. This frequency range provides a vital practical advantage compared to radio frequency microelectronics. The benefits are many, they include and not limited to improved image resolution, broader bandwidth, and sensing application. However, the terahertz frequency range also comes up with serious and major problems related to poor communication signatures, atmospheric attenuation, and general engineering challenges that adversely limit the realization of terahertz microelectronics technology into reality. Also, the functions of solid-state microelectronics operating at terahertz frequencies are hindered by the efficiency and power decline if the dimension of the active region length of the operating device is decreased to certain nano-dimensions. However, research results of diode lasers used for terahertz generation are hindered by many complications that arise from the comparison between the terahertz quantum energy and the thermal energy of the electron thus inversely populating electrons at ambient temperatures. Although considerable efforts and interest have been put in place for research and development on terahertz wave over the years, all the efforts and expectations ended up without much-desired result.

Presently, researchers around the world have shifted attention and intensified efforts in research and development for a better terahertz wave radiation that can be used in many applications like spectroscopy and image detection with a view to providing a cost-effective, and reliable terahertz sources that are more efficient and reliable than before. However, the conventional scientific problem solution stems from the need for progress in a way that cannot be used to get the solution using the existing approach. To date, the sources of terahertz waves are generally considered to come from enhanced microwave devices, resonant tunneling diodes, and or photonic devices [1–3]. Consequently, all of the traditional sources have been embedded in a series of problems such as bandwidth limitations, limited power output, operational conditions, and scope constraints. Today, the most important future promising candidate for THz wave sources comes from laser-driven sources, which operates under radiation generated by high resistivity and ultra-fast high electron mobility transistor HEMT made from III-V-based semiconductor material. Among them is the photoconductor-based material of GaAs grown at very low temperatures (LT GaAs). This is due to the materialization requirement for the deep energy levels to occur in the middle of the bandgap energy level of the semiconductor material, which led to the formation of point defects in the bandgap energy level during the growth process when molecular beam epitaxy (MBE) at low temperature is employed. However, the physics of terahertz waves involves the combinations of magnetic fields $\vec{H}$ and high-frequency electric fields $\vec{E}$ that operate from the infrared to visible ultraviolet light. The mathematical treatment of terahertz waves can be explained just like the optical waves and are essentially defined by the Maxwell equation given by Ref. [4]

$$\nabla X\ \vec{E} = -\frac{\partial}{\partial t}\ \mu\mu_0\ \vec{H} \tag{1}$$

$$\nabla X\ \vec{H} = \mathrm{J} + \frac{\partial}{\partial t}\ \varepsilon\varepsilon_0 \vec{E} \tag{2}$$

where the classical theory of electromagnetic field depends on the mathematical generalization of Maxwell's equation leading to the Helmholtz wave equation, which can also be used to treat the wave optics presented as

$$\nabla.\ \varepsilon\varepsilon_0 \vec{E} = \varrho \tag{3}$$

$$\nabla.\mu\mu_0 \vec{H} = 0 \tag{4}$$

The electrical permittivity is represented by $\varepsilon\varepsilon_0$, magnetic permeability is denoted by $\mu\mu 0$, which describes the interaction of the material with electrical and magnetic fields, respectively, while the linear approximation is represented by $\varepsilon 0$, and $\mu 0$ is the free-space values, J represents the current density while ϱ represents charge density. Maxwell's equations are the fundamental basis for the classical treatment of electrical and magnetic fields.

Since the intensity distribution for utilization of terahertz waves for sensing, medicinal, security and military is the main focus of interest, optimization and limitations of the III-V materials devices capable of operating at the frequency

Parameter	GaAs	GaAs	GaN	GaN	Si	Si
Temperature in Kelvin	300	77	300	77	300	77
Bandgap energy, E_G (eV)	1.4	1.4	3.4	3.4	1.2	1.4
Electron mobility, μ (cm^2/V s)	8000	8000	2000	2000	1440	1440
Frequency in THz	3.5	0.093	10	1.6	7.7	0.46
Breakdown field, E_C (MV/cm)	10	10	3.9	3.9	0.3	0.3

Table 1.
Material parameters and their critical frequency of operation.

range between 0.1 and 30 THz are to be presented. High electron mobility transistor (HEMT) adopts the process for modulating and coupling of terahertz wave using material surface plasmon which exists between the surface of two interfacing material layers where one surface layer on the boundary acts as negative conductive material while the other interface behaves as the positive dielectric permittivity layer. The III-V materials carrier in HEMT can be controlled or adjusted by doping, external magnetic field, or application of light temperature to demote the metallic properties into lower frequency suitable for terahertz building block devices.

Research for the III-V-based material's plasmonic properties has been conducted in the terahertz and far-infrared range by a number of groups. Furdyna and Palik's research work provided a significant insightful theory on the magneto-optical behavior of III-V materials [5]. While the work conducted by Hofmann and Shubert et al. [6, 7] describes the techniques for investigating the spectroscopic potential. Other works demonstrate the influence of terahertz time-domain spectroscopy responsible for the optical and conductive functions of III-V materials semiconductors [8, 9]. Finally, several research publications demonstrate the presence of plasmons in III-V magnetically tunable materials. **Table 1** above shows the critical frequency of operation and material properties of some III-V-based semiconductors when subjected to various degrees of temperature.

2. The III-V semiconductors

The III-V-based material semiconductors are good candidates for terahertz emission and detection applications. Because it possesses and provides higher low-field carrier mobility than silicon material transistors, which can potentially be more useful in high-quality image processing applications. The material of III-V compound semiconductors is combined with the material elements of group three and group five of the periodic table. In the crystalline of III-V elements, each atom from column III in the periodic table is bounded by four atoms of column V, while every atom of column V is bound to four atoms of column III. This means that the combination of III-V elements is bonded together by covalent bonding to form the compound of III-V semiconductors with moderate bonding ions due to differences in electro negativity that naturally exist between atoms of elements of column III and atoms of elements of column V. These characteristics provided the possibility of forming a binary, ternary or quaternary compound of semiconductors. This combination of different elements of column III and column V element into a single compound of semiconductor gives grounds for the endless possibilities in terahertz technology along with related adsorption or emission of wavelength by semiconductors material and also provides the main physical properties of teraelectronics and optoelectronics applications. Among the binary of III-V-based compound semiconductors, the most interesting materials for research investigation are GaN, AlGaN, GaAs, and their (ternary alloys).

3. Structure of the compound semiconductor

Presently, GaN and GaAs used in terahertz wave applications have been categorized into two main groups. The III-V-based (HEMTs) and the III-V-based Schottky diodes. The AlGaN/GaN HEMT structure is presented in **Figure 1**. It has three contacts: the source, the gate, and the drain. The (undoped) GaN layer is grown on either Al_2O_3 or SiC substrate. However, because of the material's difference in energy bandgap, the formation of the quantum occurs at GaN/AlGaN heterojunction. This quantum well induces a strong polarization field on the AlGaN layer and forms the most important aspect of HEMT called the 2D Electron Gas (2DEG) [10].

HEMTs rely on a heterojunction formed by two different wide bandgap materials of GaN and AlGaN or any other combination of III-V semiconductors with interfacing each other. The strain-induced polarization at the AlGaN/GaN interface produces an extraordinarily high electron mobility. **Figure 2** represents a plot showing the

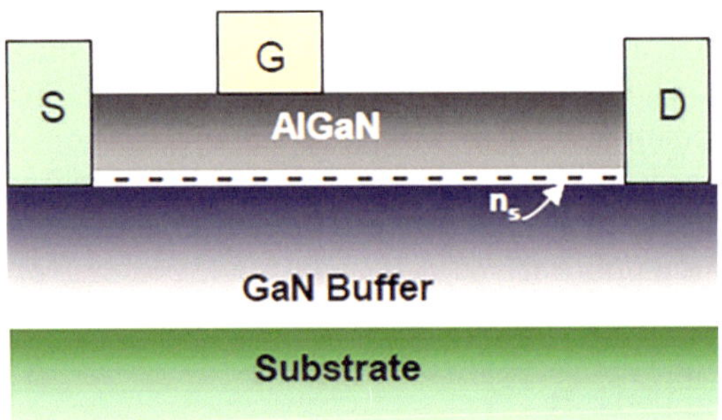

Figure 1.
Shows the typical structure of AlGaN/GaN (HEMT).

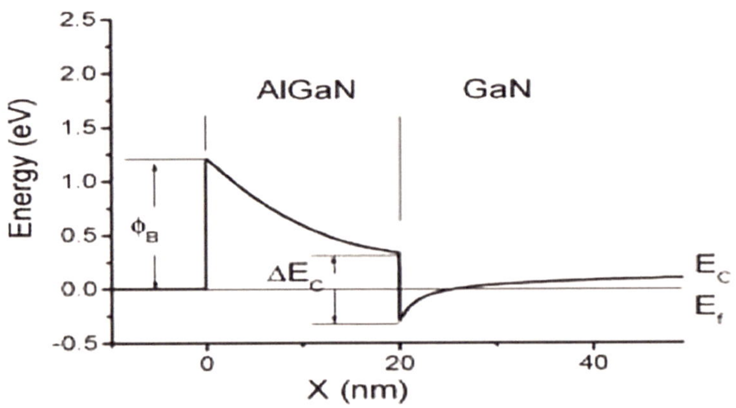

Figure 2.
Shows the energy band diagram below the Fermi level in HEMT.

bandgap bending in (HEMT). It also shows the formation of quantum well below the Fermi level at the interface of the (AlGaN/GaN) heterostructure, which confines the high mobile electrons into a narrow space. When the electrons are trapped in the quantum well, electrons would then move into the quantized energy levels within the well [11].

4. Polarization

GaN originates from a crystal structure called Wurtzite where each Gallium (Ga) atom is bonded tetrahedrally to four Nitrogen (N) atoms shown in **Figure 3**. However, this tetrahedral structure is an imperfect one due to a lack of inversion symmetry on the [0001] axis. Due to the strong electro-negative characteristics of nitrogen, changes the behavior of the GaN bond to polar and consequently leads to the formation of a higher polarization dipole (PSP) on the [0001] axis. This is called spontaneous polarization [12].

Figure 4 shows the basal-plane lattice differences between AlGaN and GaN layers. GaN is the relaxed layer below AlGaN, this creates a tensile strain within the heterostructure and produces a polarization vector on the AlGaN layer in addition to its microscopic polarization dipole. On the other hand, the corresponding negative polarization charge would be created above the (AlGaN) barrier layer as depicted in the figure. This polarization dipole in the layer of AlGaN created at the interface is what gives rise to the 2DEG [13].

The effects of the polarization can be explained in terms of charge carriers that occur as a result of the 2DEG in the channel without the need for any doping technique. The interface of the heterojunction of AlGaN/GaN and or AlInN attracts large amounts of electrons due to the positive polarization charges and confines them within the two dimensions of the quantum well [14]. The electrons can move freely only within the plane of the heterojunction and cannot go beyond the well-confined region as presented in **Figure 5**.

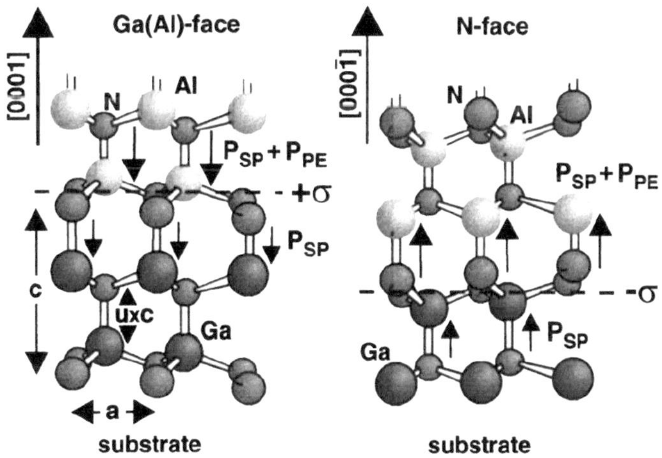

Figure 3.
Shows the wurtzite structure in GaN with the difference in spontaneous polarization along the [0001] axis due to a lack of invasion symmetry.

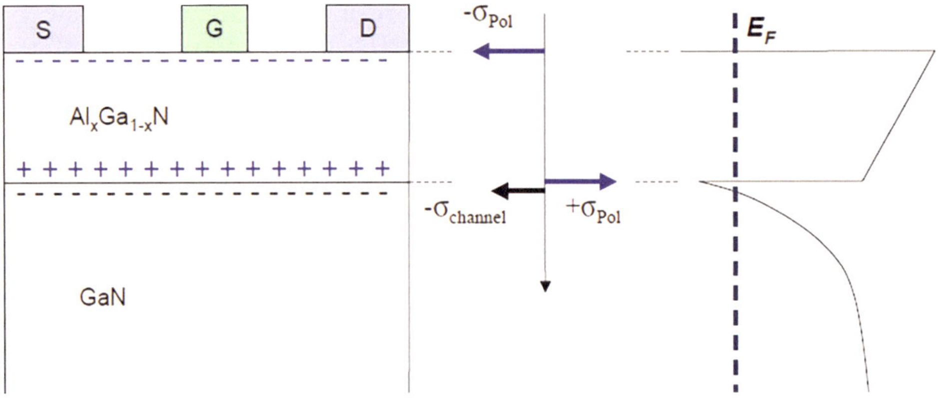

Figure 4.
Shows the basal-plane lattice-constant for AlGaN/GaN heterostructure indicating the presence of tensile strain on the surface of the AlGaN layer.

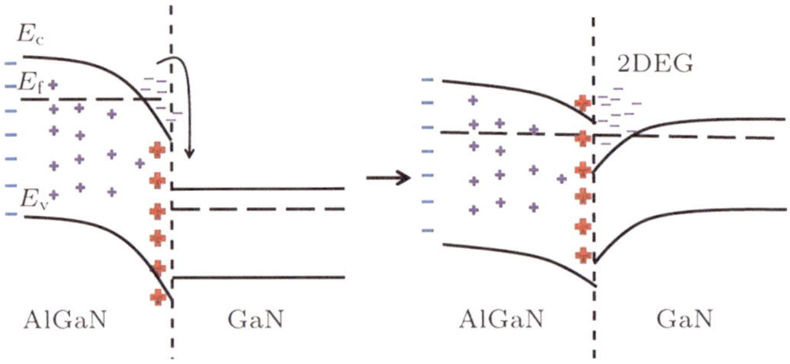

Figure 5.
The different Fermi levels created by the energy band at the interface of AlGaN/GaN provide an electron flow into the channel shown on the left and the entire band energy heterostructure creates the 2DEG within the quantum well shown to the right.

5. Formation of the two-dimensional-electronic gas

A Band diagram of AlGaN/GaN heterostructure indicating the various stages of charge components is presented in **Figure 6**. The electrostatics of the device across the heterostructure are responsible for quantum well realization on both interfaces within which the 2DEG is formed. It is possible for a truly undoped AlGaN barrier where its upper surface acts as the donor and becomes the source of electrons of the 2DEG, which is due to charge balance across the heterostructure.

The formation of strong 2DEG in the AlGaN barrier layer due to polarization dipole is presented. The barrier thickness, which increases as the Fermi level decreases, ionizes on both smooth surfaces and traps a huge amount of electrons in the quantum well resulting in the formation of the 2DEG [15].

6. The III-V-based terahertz electronics

For many decades, high electron mobility transistors (HEMT) have been in research focus for possible applications in spectroscopy and terahertz wave

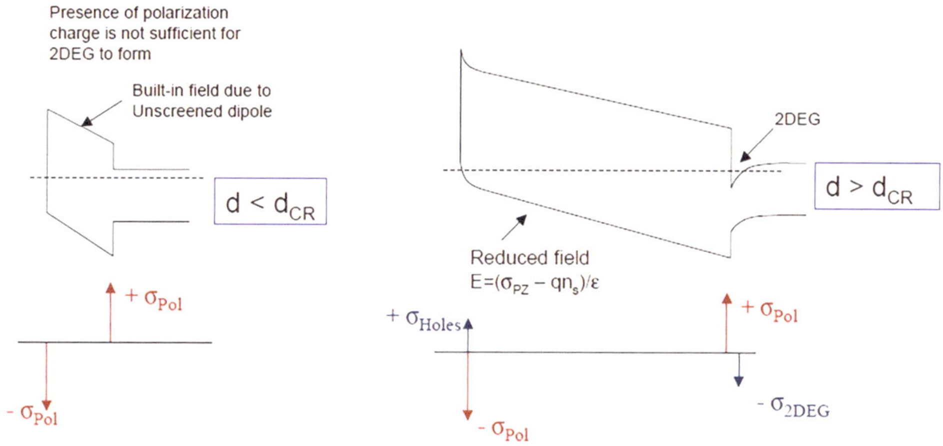

Figure 6.
Showing the band diagrams at various charge components where the critical thickness depends on bandgap and polarization is not directly responsible for the 2DEG at the interface.

application. This in part is due to the presence of active plasma excitation caused by the 2DEG. However, the physics behind the terahertz wave emission, detection, and sensing by III-V-based HEMT embroils quasi-ballistics, collision-dominated, and viscosity-controlled routines for hole or electron transports due to the linear and non-linear physical properties. Where both electron and hole transportation carrier occur at distinctive sizes with impurity and the phonon scattering became prevalent due to the quasi-ballistic nature of the waves. Terahertz radiation generally depends on the electric field scattering scale available in the device. This physical property makes it an exceptional candidate for high-resolution scanning applications. For the short channel field effect transistors of different material sizes such as InAs (0.9–3.5 μm), InGaAs (1.1–1.7 μm), and Ge (0.7–1.8 μm) along with other 2D materials are found to be suitable to operate within the terahertz frequencies that range from 0.1 to 5 THz.

This can be explained using the conductive and optical properties of the semiconductors operating within the infrared and terahertz spectrum. These properties are dependent on their lattice vibration, free carrier absorption, and background permittivity, which are represented using the Drude-Lorentz function given as [16].

$$\varepsilon_r = \varepsilon_\infty - \frac{\omega_p^2}{\omega^2 + i\gamma p\omega} + \frac{A_L\ \omega_L^2}{\omega_L^2 - \omega^2 - \omega_p^2} \tag{5}$$

where ε_∞ is the background permittivity of the semiconductor, the Drude term ω_p originating from the free carrier is given by the following relation

$$\omega_p = \left(\frac{Ne^2}{\varepsilon o m^*}\right)^{1/2} \tag{6}$$

N represents the carrier concentration, ε0 represents permittivity, and electron charge is represented by e, m∗ represents a mass of charge carriers, γp represents the damping constant. Lorentz term ε_L is the factor that describes lattice vibration of the frequency ω_L of the damped semiconductor. **Figure 7** shows the terahertz band. The terahertz region is the region that lies between 0.1 and 10 THz and separates electronics from photonics, and has over the years shown to be the region that is

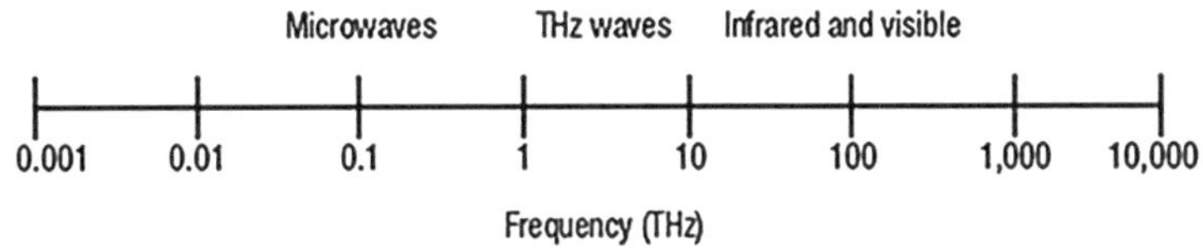

Figure 7.
Shows the frequency spectrum indicating the position of the terahertz band.

difficult to access. This new development has provided the avenue for extending the electronic range into terahertz.

Research findings demonstrated that infrared optoelectronics is unsuitable to operate at the high-end frequency of 10 THz in the spectrum due to the photon-electron particle interactions employed which significantly decreases the photon energy $h\nu$ below the device thermal energy kT and ceases to efficiently operate at ambient temperature. This makes the operating device require an additional cooling system that adds up extra weight and cost. The electron diffusion speed across the semiconductor heterostructure decreases consequentially increasing the transport time [17].

When the GaAs layer interfaces $Al_xGa1_{-z}As$ barrier layer, electron relaxation occurs between the two interface layers resulting in the serially connected sub-bands of the quantum well generating and emitting the terahertz waves. However, there are three main approaches usually adopted for developing a terahertz wave using this method. These are the interlaced, the bound to continuum, and the resonant phonons [18]. The three differ from each other by way of electron emission and scattering originating from the source to the target leading to the surge of power increase to tens of milliwatts. The small system complexity, high sensitivity at ambient temperature, and fast time response offered by GaAs-based devices gave it an edge over other materials [19]. Dyakonov et al. [20] and Shur et al. [21] have proposed a mathematical model to back up the TeraFETs theories. The proposed mathematical model demonstrates how a significant concentration of 2DEG (two-dimensional electron gas) presence in the channel can produce a photovoltage reaction in the form of direct current (DC), which produce terahertz radiation. The model demonstrates the wave excitation of the plasma carriers in the channel give rise to rectilinear properties.

The full electromagnetic waves produced by the two-dimensional material devices of passive circuit components have been demonstrated on graphene absorbers using simulation. The techniques successfully cover the whole terahertz frequency range [22, 23]. When compared to different kinds of other transistors, GaN HEMTS demonstrates excellent power and frequency characteristics. The excellent amount of 2DEG concentration in the channel formed by the spontaneous polarization enhances the saturation velocity without the need for any intentional doping [24].

GaN HEMT-based ultra-wideband terahertz detectors that cover the frequency range of 0.2–1.0 THz have been successfully simulated with excellent results [25]. The detectors use a waveguide coupling approach and demonstrate a noise equivalent to power of 190 pW/Hz [26]. The approach further enhances the waveguide to attain 76 pW/Hz at 227 GHz [27]. Other approaches designed with a combination of 18 dipole antennas successfully achieved a noise that corresponds to the power equivalence of 188 pW/Hz at a frequency that ranges from 0.2 to 1.1 THz at room temperature. Also, another approach was used to detect the terahertz wave by Maris

Bauer involving a plasma wave-based detection principle of antenna factor using a field effect transistor and operating the antenna within the broadband spectrum at the high end of the frequency [28]. Jiandong Sun et al. also proposed an improvement of the sensitivity of the detector by way of decreasing the gate length and the space between the gate and the antenna [29].

Accordingly, another new model of recessed graded gated AlGaN HEMT has been proposed to improve the electrical properties of HEMT in relation to terahertz response to the detector [30]. The simulation works demonstrate the presence of low channel resistance in GaN HEMT with a high field effect factor can further be used to enhance the detector with the same coupling structure. The work employs the graded barrier layer of AlGaN to generate 3DEG (Three-dimensional electron gas), which reduces the channel resistance and successfully achieves a high field effect factor with low channel resistance, resulting in a significant improvement of the device sensitivity. When the electrical characteristics of the detector were examined, it showed that the new structure enhances the current degradation, which affects the conventional structure. The result shows that this new proposed structure is a promising candidate for improving the performance of the detector.

7. Structure and electrical characteristics GaAS (LT GaAS)

Excess high-density correlated defects observed in the low-temperature GaAS were the reason behind the high breakdown fields found in GaAS (LTG). The result shows that when GaAs are grown at the temperature of 200C, their electric field changes to be as higher as 320 kV/cm with a small reduction when annealed to ~250 kV/cm at the temperature of 650°C and remains high in one order of magnitude above other III-V semiconductors [31]. This shows that the observed improvement of the breakdown field in GaAS at low temperatures is attributed to Si-based substrate which eliminates the surface breakdown, secondly, the high conduction observed reduces the accumulation of much-injected electrons on the surface and prevents early breakdown.

8. Some application of terahertz

The principles upon which terahertz technology is based are classified into two main categories; these are communication and sensing. The terahertz Frequency Domain Spectroscopy (FDS) and Time Domain Spectroscopy (TDS) are the current innovative technological tools applicable to many materials. Application of the tools gives extraordinary sensing abilities in many scientific and technological fields. This is not limited to DNA chips, very large-scale integrated (VLSI), explosive inspection and cancer diagnostic among others. Other fields such as information and communication technology in satellite communication, for example, high-speed data processing and wireless communication.

8.1 The terahertz image detection

Terahertz wave emission for use in image detection is another recent technology that unlocks many opportunities in the areas of military, medical, manufacturing,

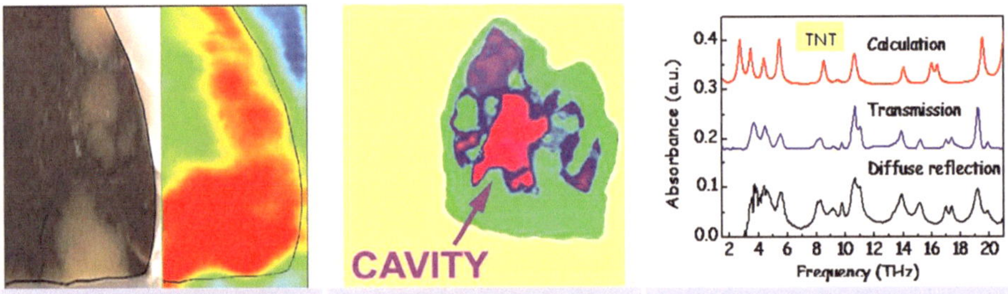

THz cancer detection.[1] | THz applications in dentistry.[2] | Explosive detection using THz radiation.[3]

Figure 8.
Shows applications of terahertz waves for image detection used in medical and security fields. From left to right: Cancer detection in the human body, spot tooth cavity earlier than x-ray, and explosive detection using THz radiation.

and security fields. Terahertz waves just like infrared and other radio waves can be used to penetrate any material to assess the concealed material using short wavelengths to capture images with high resolution. Terahertz wave is also preferred in the medical field over X-ray imaging due to its advantage of not having any harmful effect of high-energy radiation. The terahertz-captured image can be recorded in real-time without the need for the use of any film like in X-ray to process and evaluate the patient in real-time. In medical applications, for example, terahertz is used for teeth and cancerous cell and tissue screening using image samples as shown in **Figures 8** and **9** [32].

Additionally, military and security outfits that periodically screen people and luggage in search of explosives, weapons, or contraband adopt the use of terahertz image systems. Other arrears include quality control, electronic circuit wire bonding, and inspection of manufactured goods in the manufacturing fields.

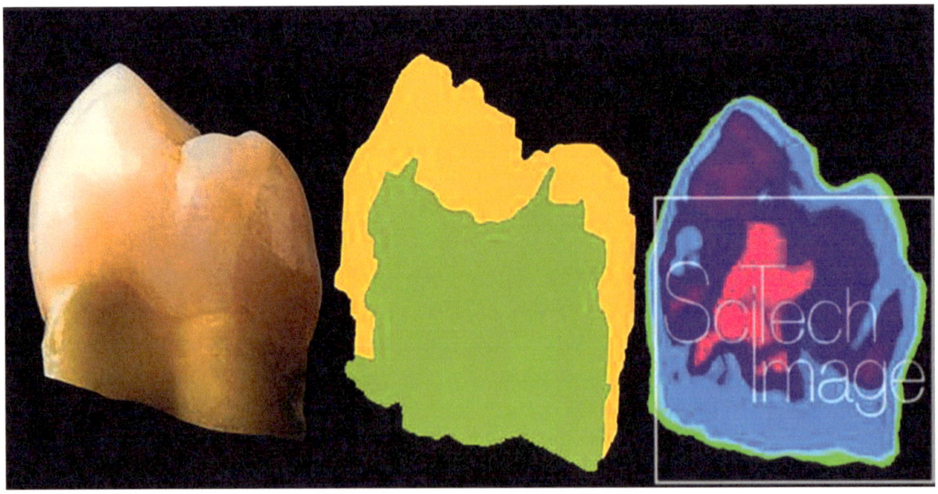

Figure 9.
Shows applications of THz waves in dentistry. From left to right; image of tooth using the visible light camera, terahertz image where the yellow color represents the enamel of the dentine area as green, and the left represents cavity identified in pink by terahertz absorption.

8.2 Security

Narcotic and explosive detection technology uses terahertz spectra to identify a distinct signature by using terahertz spectroscopy. The signature is used to identify chemicals using their transmission spectra, which distinguish explosive and illegal drugs. Other applications of terahertz wave by THz camera are notably at airport security. The camera uses silicon photonic-based crystal with heterodyne techniques operating at a frequency not lower than 500 GHz for detection. The camera can also scan images of objects including biometrics and through smoke or fog irrespective of background light. Terahertz waves are actively used to produce femtosecond laser beams, which generate ambient plasma where the reflection of the target can be detected. Recently, terahertz sensing was used to detect hazardous gas.

9. Conclusion

Investigation of Ill-V-based compound materials used for terahertz and opto-electronic frequency detectors and emitters have been reviewed and presented. It is reported that field application of homojunction and heterojunction of HEMT requires material with exceptional parameters of high resistivity, high electron mobility, and ultra-short electron generation to emit or detect terahertz waves. A new set of parameters was discovered when the material of GaAs was grown at low temperatures using the MBE technique. The recessed-gated III-V-semiconductor-based high electron mobility transistor was activated to release pulses in the terahertz region.

Author details

Yusuf U. Tarauni[1]* and John Thiruvadigal[2]

1 Department of Physics, Aliko Dangote University of Science and Technology ADUSTECH, Kano, Nigeria

2 Department of Physics and Nanotechnology, SRM Institute of Science and Technology, Chennai, India

*Address all correspondence to: yu4497@srmist.edu.in

IntechOpen

© 2025 The Author(s). Licensee IntechOpen. This chapter is distributed under the terms of the Creative Commons Attribution License (http://creativecommons.org/licenses/by/4.0), which permits unrestricted use, distribution, and reproduction in any medium, provided the original work is properly cited.

References

[1] Eisele H. Recent advances in the performance of InP Gunn devices and GaAs TUNNETT diodes for the 100-300 GHz frequency range and above. IEEE Transactions on Microwave Theory and Techniques. 2000;**48**:626

[2] Fong T, Kuno H. Millimeter-wave pulsed IMPATT sources. IEEE Transactions on Microwave Theory and Techniques. 1979;**27**:492

[3] Brown E, Soderstrom J, Parker C, Mahoney L, Molvar K, McGill T. Oscillations up to 712GHz in InAs/AlSb resonant-tunneling diodes. Applied Physics Letters. 1991;**58**(20):2291

[4] Piprek J. Semiconductor Optoelectronic Devices Introduction to Physics and Simulation. USA: Academic Press an Imprint of Elsevier Science; 2003

[5] Ferguson B, Zhang X-C. Materials for terahertz science and technology. Nature Materials. 2002;**1**:26-33

[6] Hofmann T, Herzinger CM, Krahmer C, Streubel K, Schubert M. The optical hall effect. Physica Status Solidi A. 2008;**205**:779-783

[7] Schubert M, Hofmann T, Šik J. Long-wavelength interface modes in semiconductor layer structures. Physical Review B. 2005;**71**:035324

[8] Mittleman DM, Cunningham J, Nuss MC, Geva M. Noncontact semiconductor wafer characterization with the terahertz hall effect. Applied Physics Letters. 1997;**71**:16

[9] Kadlec F, Kadlec C, Kužel P. Contrast in terahertz conductivity of phase-change materials. Solid State Communications. 2012;**152**:852-855

[10] Hashimoto S, Akita K, Yamamoto Y, Ueno M, Nakamura T, Takeda K, et al. High carrier concentration in high Al-composition AlGaN-channnel HEMTs. Physics Status Solidi C. 2012;**9**(2):373-376

[11] Hashimoto S, Akita K, Yamamoto Y, Ueno M, Nakamura T, Takeda K, et al. Enhancement of two-dimensional electron gases in AlGaN-channel high-electron-mobility transistors with AlN barrier layers. Physica Status Solidi A. 2012;**209**(3):501-504

[12] Baca AG, Armstrong AM, Allerman AA, Douglas EA, Sanchez CA, King MP, et al. An AlN/$Al_{0.85}Ga_{0.15}N$ high electron mobility transistor. Applied Physics Letter. 2016;**109**(3):033509-1-033509-4

[13] Anbuselvan N. "Analytical modeling of 2DEG and 2DHG charge balancing in quaternary" Al0.42In0.03Ga0.55N/Al0.3In0.7N HEMTs. Journal of Computational Electronics. 2018;**17**:1191-1198

[14] Tarauni YU, Thiruvadigal J, Joseph B. Characterization and optimization of MIS-HEMTs device of high k dielectric material on quaternary barrier of $Al_{0.42}In_{0.03}Ga_{0.55}N$/UID-AIN/GaN/GaN heterostructure for high power switching application. Applied Surface Science. 2019;**488**:427-433. DOI: 10.1016/j.apsusc.2019.05.170

[15] Ambacher O, Smart J, Shealy JR, et al. Two-dimensional electron gases induced by spontaneous and piezoelectric polarization charges in N-and Ga-face AlGaN/GaN heterostructures. Journal of Applied Physics. 1999;**85**(6):3222

[16] Berreman DW. Optics in stratified and anisotropic media: 4 × 4 -matrix

formulation. Journal of the Optical Society of America A. 1972;**62**:502-510

[17] A New Technology for Terahertz Electronics, Thin Film Devices Based on Metal-Insulator Tunnel Junction. Based on a white paper for Phiar corporation – 13th August 2003

[18] Toktam Aghaee, Ali A. Orouji, Reconfigurable multi-band, graphene-based THz absorber: Circuit model approach, Results in Physics. 2020;**16**:102855. DOI: 10.1016/j.rinp.2019.102855

[19] Yusuf U. Tarauni J, Thiruvadigal BJ, Mohan B. Optimization of enhancement mode P-type Mg-doped In0.2Ga0.8N cap gate DH-HEMTs for low-loss high power efficient boost converter circuits, Material Science in Semiconductor Processing. 2019;**103**:104624. DOI: 10.1016/j.mssp.2019.104624

[20] Dyakonov M. Detection, mixing, and frequency multiplication of terahertz radiation by two-dimensional electronic fluid. IEEE Transactions on Electron Devices. 1996;**43**(3):380-387

[21] Sun J. Probing and modelling the localized self-mixing in a GaN/AlGaN field-effect terahertz detector. Applied Physics Letters. 2012;**100**:173513

[22] Vicarelli L. Graphene field-effect transistors as room-temperature terahertz detectors. Nature Materials. 2012;**11**(10):865-871

[23] Bianco F. Terahertz detection by epitaxial-graphene field-effect-transistors on silicon carbide. Applied Physics Letters. 2015;**107**:131104

[24] Tarauni YU, Thiruvadigal J, Hotoro M, Koki FS, Murugapandiyan P. Influence analysis of back-barrier and AIN substrate on the high-temperature performance of an E-mode Mg-doped $In_{0.2}Ga_{0.8}N$ capped gate high electron mobility transistor for high-power switching applications: A simulation study. Journal of Electronic Materials. 2022;**51**:5219-5229. DOI: 10.1007/s11664-022-09767-5

[25] Xie Y. Defect engineering of MoS(2) for room-temperature terahertz photodetection. ACS Applied Materials & Interfaces. 2020;**12**(6):7351-7357

[26] Rashidi B, Soldoozy A, Mohammad AA. Switchable terahertz wave absorber based on disks and its complement graphene surfaces. Results in Optics. 2024;**16**:100719

[27] Sirjani S. THz wave manipulation via multiple bias circuits based on combined graphene patterns. Indian Journal of Physics. 2024;**98**:339-348

[28] Hou H. A sub-terahertz broadband detector based on a GaN high-electron-mobility transistor with nanoantennas. Applied Physics Express. 2017;**10**:014101

[29] Cen C. Numerical investigation of a tunable metamaterial perfect absorber consisting of two-intersecting graphene nanoring arrays. Physics Letters A. 2019;**383**:3030-3035

[30] Li Y, Sun J, Qin C, Deng X, Zheng X. A high- performance terahertz detector based on a graded channel GaN high-electron-mobility transistor with a recessed gate. Optical Materials. 2024;**150**:115233

[31] Luo JK, Thomas H, Morgan DV, Westwood D. Thermal annealing effect on low temperature molecular beam epitaxy grown GaAs: Arsenic precipitation and the change of resistivity. Journal of Applied Physics. 1996;**76**:3622

[32] Jefferson Lab. Available from: http://www.lightsources.org/cms/?pid=1000680

Chapter 3

Antennas in Terahertz Technology

Somdotta Roy Choudhury

Abstract

The rapid advancement in wireless technology over recent years has significantly transformed data traffic patterns. Terahertz (THz) frequencies can enable data transfer rates far exceeding those possible with current technologies (such as Wi-Fi and 5G). The potential for multi-gigabit-per-second transfer rates opens up new possibilities for applications requiring high bandwidth. The shorter wavelengths of terahertz signals allow for more compact antennas and enhanced spatial resolution. This can lead to smaller and more efficient wireless devices that can operate within densely populated environments. Using THz frequencies could alleviate congestion in existing communication bands, offering a vast spectrum of unused frequencies that can be utilized for data transmission, especially in urban areas with high traffic. THz technology has the potential to impact a variety of sectors, including: healthcare: non-invasive imaging and diagnostic techniques; telecommunications: ultra-high-speed internet and wireless backhaul solutions; consumer electronics: improved Wi-Fi standards, augmented reality, and virtual reality applications. Various research progress has been made for different types of THz antennas, including photoconductive antennas, horn antennas, lens antennas, microstrip antennas, on-chip antennas, etc.

Keywords: antennas, terahertz gap, wideband, multiband, metasurface

1. Introduction

Terahertz (THz) antennas are designed to transmit and receive electromagnetic waves in the terahertz frequency range (from 0.1 to 10 THz). These frequencies span between the microwave and infrared regions of the electromagnetic spectrum. The interest in THz technology has surged due to its potential applications in various fields, including communications, imaging, spectroscopy, and medical diagnostics. These antennas are of small size, wide frequency bandwidth, and high data rate. Terahertz (THz) antennas face significant challenges related to high losses and low fabrication precision. At THz frequencies, materials can exhibit high intrinsic losses. Conductors may not perform as well due to skin effect, which causes the currents to flow mainly at the surface, leading to increased resistance. Similarly, dielectric materials can absorb THz radiation, which leads to additional losses in the system. The efficiency of radiation can be affected by the design and size of the antenna. Nano-scale dimensions compared to the wavelength can lead to inefficient radiation patterns.

The small dimensions of THz antennas shown in **Figure 1** (often in the micrometer to nanometer range) require high-precision fabrication techniques. Traditional lithography and etching methods may not achieve the required tolerances

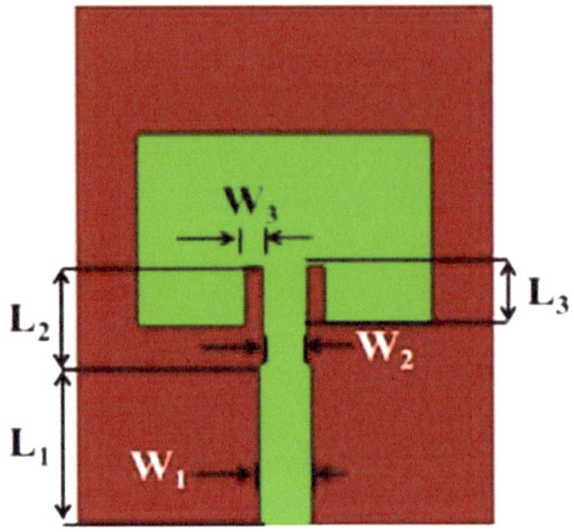

Figure 1.
Prototype of THz patch antenna [1].

consistently. Many materials used for antenna fabrication at microwave or RF frequencies might not translate well to THz applications, necessitating the development of new materials or composites that can be more challenging to manufacture. Integrating THz antennas with other devices (e.g., sensors, detectors) adds complexity and can affect performance due to misalignment or mismatch in material properties. Terahertz communication technology has demonstrated the potential for achieving gigabit per second (Gbps) data rates, and work is ongoing to explore and push the boundaries toward terabit per second (Tbps) data rates. THz wave is defined at 0.3–10 THz by IEEE standard. Terahertz (THz) waves [1–5], as shown in **Figure 2**, which lie in the electromagnetic spectrum between microwave and infrared frequencies, exhibit several excellent characteristics, making them valuable for various applications in science and technology.

Some of these characteristics include:

1. High penetration depth: THz waves can penetrate non-conductive materials, such as plastics, fabrics, and biological tissues, which allows for non-destructive testing and imaging.

2. High sensitivity: THz spectroscopy can detect small changes in material properties and is sensitive to various molecular vibrations, making it useful for identifying substances and studying chemical processes.

3. Real-time imaging: THz technology can provide real-time imaging capabilities, which is beneficial in sectors like medical imaging and security screening.

4. Low energy levels: THz waves have relatively low energy compared to X-rays, making them safer for biological applications and reducing the risk of ionizing damage to tissues.

5. Spectroscopic insight: They are effective for spectroscopy, allowing the study of molecular structure, dynamics, and interactions due to their ability to probe vibrational and rotational transitions in molecules.

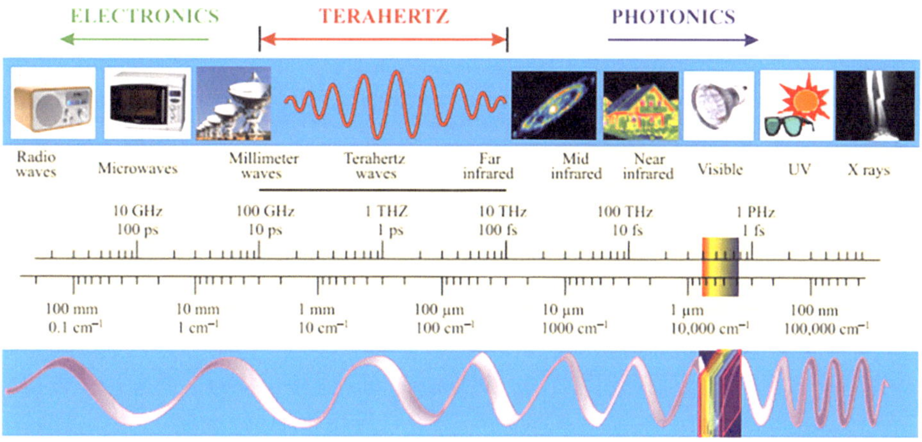

Figure 2.
THz wave in the electromagnetic spectrum [2].

6. Non-ionizing radiation: THz radiation does not ionize atoms or molecules, which makes it safer for a variety of applications, including medical imaging and security scanning.

7. Versatility: THz technology is versatile, finding applications in areas such as telecommunications, imaging systems, material characterization, and security.

8. Small device size: Advances in technology have led to the miniaturization of THz sources and detectors, making them suitable for portable and compact applications.

9. Potential for data transmission: THz waves have the potential to support high data transmission rates, making them a candidate for future communication systems, such as wireless networks.

10. Imaging and sensing applications: THz technology can distinguish between different materials and can be used for quality control, art restoration, and biological research.

While terahertz (THz) antennas [5–9] and technologies are still in the developmental stages compared to other frequency ranges, there are several notable applications and potential use cases that are emerging, such as imaging and sensing, high-speed wireless communication, material characterization, wireless networks, satellite communication, various medical applications, etc., shown in **Figure 3**. Antennas operating in the terahertz (THz) band indeed offer significantly higher bandwidth compared to traditional antennas that operate in lower frequency bands. THz band provides a wider usable frequency band compared to millimeter waves. It also exhibits some distinct characteristics when compared to light waves, especially in terms of penetration abilities and efficiency for certain applications. One of the standout features of THz waves is their wide operating bandwidth, which is a significant advantage for THz antennas. This ability to operate over a large frequency range enables THz antennas to support high data rates and enhance resolution in imaging

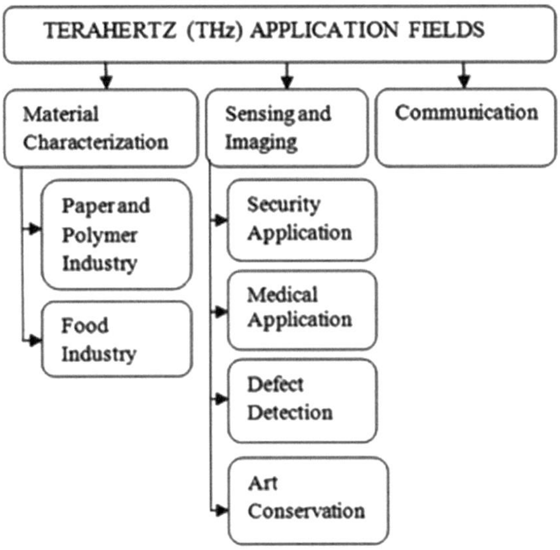

Figure 3.
Possible terahertz technology applications [3].

applications. THz antennas possess several advantageous characteristics such as wide frequency bands, high resolution, strong directivity, and miniaturization. These antennas face several challenges, like fabrication precision required for reduced size, material limitations, atmospheric absorption, advanced signal processing techniques to handle the high data rates, and higher costs compared to microwave technologies. The development of terahertz (THz) antennas represents a significant advancement in the field of antenna technology [10–12].

In this chapter, the first basic design and development of terahertz antenna are discussed in Section 2. Then in Section 3, different types of antennas in terahertz frequencies, such as THz Horn antenna, THz Lens antenna, and THz microstrip antennas, are described in detail. The challenges regarding THz antennas are depicted in Section 4. The future work and conclusions are discussed in Section 6, respectively.

2. Design and development of THz antenna

The development of wireless communication technology at the beginning of the twenty-first century has been remarkable for THz antennas. The growing demand for information and advancements in communication technology have significantly influenced the requirements for data transmission rates. Microwave antennas typically operate in frequency bands below 30 GHz, which have limited bandwidth. They are suitable for moderate data rates but struggle with the higher throughput demands of modern applications. Again, Millimeter-Wave antennas operate in the frequency range of 30–300 GHz and can support significantly higher data rates due to wider bandwidth availability. This makes mmWave suitable for applications like 5G and beyond, where large data flows are frequent. In 5G and beyond, the user data connection speeds exceed Gigabits per second (Gbps), which would indeed have a significant impact on data traffic for base stations. In this scenario, when it comes to handling the increasing demand for data transmission that is driven by high-definition video streaming, virtual reality, and the growing number of connected devices, traditional microwave links may struggle to keep up for several reasons, like bandwidth limitations, atmospheric absorption, rain fade, and obstacles like buildings, etc. The higher frequencies in THz bands enable much greater bandwidth, which can facilitate faster data transmission rates. In comparison to microwaves, THz waves can support more data channels, leading to enhanced capacity for information transfer. Therefore, THz frequencies for ultra-high-speed wireless communication systems is a promising approach to address the growing demand for high data rates. These systems can potentially support gigabit and even terabit data transfers, making them ideal for various aforementioned applications shown in **Figure 4**. The establishment of a THz communication system operating at 0.12 THz in 2004 marked a significant milestone in the field. This achievement demonstrated the feasibility of using THz frequencies for wireless communication [13], opening the door to a range of applications due to the high bandwidth and data rates associated with these frequencies. Antenna design and configuration in THz frequency evolved in 2004, and the whole communication system matured in 2013. Nippon Telegraph and Telephone (NTT) Corporation developed two types of THz antennas, one is an indoor planar antenna with a uni-traveling carrier photodiode (UTC-PD) chip, a planar slot antenna chip, and a silicon lens, and the other one is an outdoor THz antenna with a Gaussian optical lens, a feed horn, and a dielectric lens. NTT's development of a 0.3 THz communication system in 2012 marked a significant advancement in high-frequency wireless communication technology. In Japan, terahertz (THz) communication systems have demonstrated impressive data transfer rates, achieving speeds of up to 10 Gbps in 2004. By combining THz communication with photonic technologies, researchers can enhance the generation, transmission, and reception of THz signals in 2006. The integration of BPSK modulators and demodulators on a Microwave Monolithic Integrated Circuit (MMIC) platform was a significant achievement at THz frequency in 2009. In 2013, terahertz (THz) communications can significantly enhance data transmission rates, allowing for speeds like 50 and 100 Gbps by improving the baseband circuit

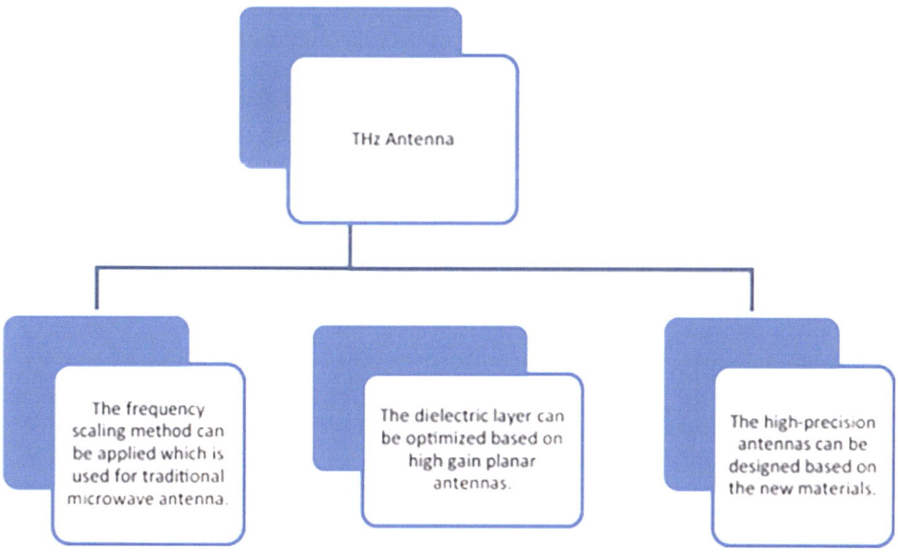

Figure 4.
Design concepts of THz antennas [5].

bandwidth of the detector. In a terahertz (THz) communication system, modeling a causal channel developed in 2018 [14], which incorporates the effects of molecular absorption, is crucial for understanding signal propagation and performance. The hybrid approach of combining wireless and fiber links facilitates high-capacity data transmission, such as 8 Gbps error-free transmission and uncompressed HD 4 K video, introduced in 2019, which provides high bandwidth and low latency over long distances, while wireless links offer flexibility and ease of deployment. THz all-dielectric rod antenna arrays are an exciting advancement in high-frequency communications. Achieving a 28% relative bandwidth with over 20 dBi gain is impressive, as it allows for efficient signal transmission and reception across a wide range of frequencies. Optimizing THz antennas is essential for maximizing the performance of THz communication systems. By focusing on materials, design, integration, and testing, the efficiency, bandwidth, and overall system capabilities can be significantly enhanced, paving the way for advanced applications in telecommunications, sensing, and beyond [15–17]. The extensive research conducted in the U.S. on THz wave emission and detection reflects the technology's potential to revolutionize multiple fields. Research into THz communication aims to leverage high data rates for wireless networks, addressing the growing demand for bandwidth. Following laboratories and institutions like Jet Propulsion Laboratory (JPL), Stanford Linear Accelerator Center (SLAC), National Aeronautics and Space Administration (NASA), National Science Foundation (NSF), etc., are pivotal in advancing THz research, driving innovations that have implications in telecommunications, security, materials science, and beyond [12, 18–21].

Their collaborative efforts continue to push the boundaries of what is possible with THz technology. The development of terahertz (THz) antennas has led to several innovative design concepts. Here are three basic design ideas commonly used in current THz antenna technologies, such as THz antennas that maintain the beneficial characteristics of their microwave counterparts while operating efficiently in the THz range. Another type is high-gain planar THz antennas with optimized dielectric layers, which give improved efficiency, bandwidth, and directivity, making them suitable for advanced applications in communication, imaging, and sensing. Third type is based on the development of terahertz (THz) antennas using new materials, which leads to high-precision designs that enhance performance in various applications. One of the main challenges of THz antennas is narrow bandwidths, which can restrict their ability to operate effectively across a range of frequencies. This is particularly problematic for applications requiring broad-spectrum operation. At THz frequencies, many materials exhibit high dielectric losses, which can significantly attenuate the signal. This limits the efficiency of antennas and can reduce their effective range and sensitivity. Again, the small dimensions and tolerances needed at THz frequencies make manufacturing more challenging than at lower frequencies. Once more, THz waves are sensitive to environmental factors like humidity and atmospheric absorption, which can lead to signal degradation. This sensitivity can limit their effective range, particularly in outdoor applications [22, 23]. The existing research on terahertz (THz) antennas primarily concentrates on several key areas:

- Material development: Researchers are exploring new materials, such as 2D materials (like graphene) and metamaterials, to improve the performance and efficiency of THz antennas. The focus is on minimizing losses and enhancing sensitivity.

- Antenna design: Innovative designs are being developed, including planar, patch, and nano-structured antennas. These designs aim to optimize radiation patterns, gain, and bandwidth, addressing the unique challenges of THz frequencies.

- Fabrication techniques: Advanced fabrication methods, such as 3D printing, nanoimprinting, and lithography, are being researched to achieve precise geometries and improve manufacturing scalability.

- Integration with THz sources and detectors: There is a strong emphasis on integrating THz antennas with efficient sources and detectors to enhance overall system performance. This includes studying coupling techniques and impedance matching.

- Applications in imaging and sensing: Much research is directed toward practical applications of THz antennas in fields like biomedical imaging, security screening, and environmental monitoring, exploring how they can improve existing technologies.

- Signal processing and communication: Research is being conducted on the use of THz antennas in communication systems, focusing on their potential for high-data-rate transmission and the development of THz communication protocols.

- Environmental effects: Understanding the impact of atmospheric conditions and other environmental factors on THz signal propagation is another important area of study, aiming to improve the robustness of THz systems.

- Theoretical modeling and simulation: Developing accurate models and simulation tools to predict the performance of THz antennas under various conditions is crucial for guiding design and optimization efforts.

3. Types of terahertz antennas

Basic terahertz (THz) antennas can be categorized into several types, each with distinct designs and characteristics.

Dipole antennas: These are widely used due to their simplicity and efficiency in coupling THz waves. They are simple and straightforward, consisting of two conductive elements and can be designed for specific bandwidths (**Figure 5**).

Microstrip antennas: These antennas utilize a thin dielectric substrate with a conductive patch to create terahertz radiation. They can be designed with various shapes to optimize performance.

Plasmonic antennas: These utilize surface plasmons (coherent delocalized electron oscillations that exist at the interface between metals and dielectrics) to enhance the interaction with THz waves.

One of the different available types of THz antennas is a pyramidal cavity with a dipole antenna, which is an interesting configuration that combines the properties of both structures to enhance performance for terahertz applications. Again, an angle reflector array THz antenna offers significant advantages in terms of gain, directivity, and application versatility, making it a valuable design choice for advancing THz technologies. The bow-tie dipole THz antenna is a versatile and effective design

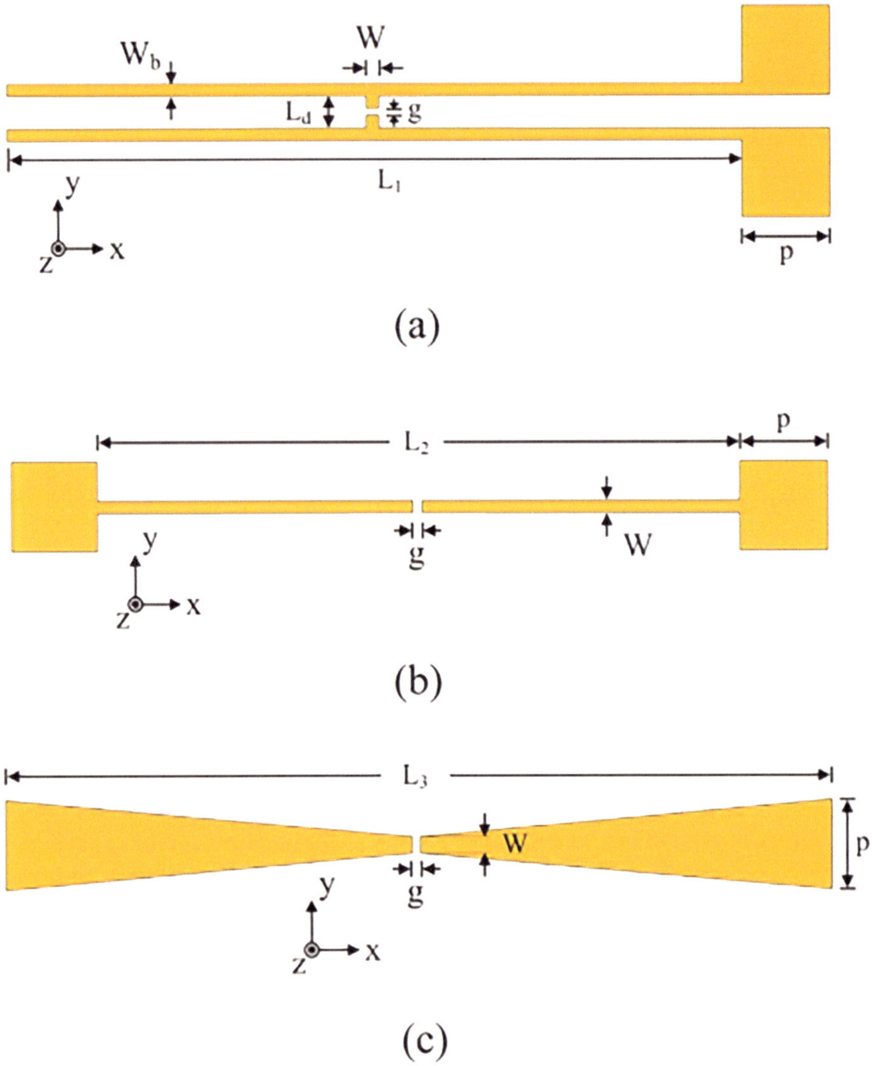

Figure 5.
Different types of THz dipole antennas [24].

suitable for various THz applications. The dielectric lens planar THz antennas represent a powerful approach to enhancing the capabilities of THz systems. These antennas combine the principles of lens design with planar antennas. Photoconductive antennas utilize the photoconductive effect to convert laser light into THz electromagnetic waves. Graphene can be incorporated into metamaterial designs to create advanced antennas with tailored electromagnetic properties, further enhancing THz performance.

Again, THz antennas can be classified based on their manufacturing materials into several categories. Like:

- Metallic antennas: dipole, patch, and horn antennas made of copper, aluminum, and gold. These antennas suffer from losses at THz frequencies due to skin effect.

- Dielectric antennas: These antennas are made of dielectric materials, including polymers (like PTFE) and ceramics. Dielectric antennas are often used in conjunction with metallic elements to reduce losses and improve performance.

- Graphene-based antennas: Graphene antennas can be highly tunable and responsive, making them suitable for broadband applications. Their two-dimensional nature allows for compact designs.

- Metamaterials: These antennas are made of combinations of different materials engineered to achieve specific electromagnetic properties. Metamaterial-based antennas can exhibit unusual properties such as negative refractive index, which can enhance performance.

- Plasmonic antennas: Typically made from metals such as gold or silver that support surface plasmons. These antennas leverage plasmonic effects to enhance sensitivity and performance at THz frequencies, making them ideal for applications in sensing and imaging.

- Hybrid antennas: These antennas are made of combinations of the above materials, such as metallic-dielectric or metallic-graphene structures. Hybrid designs aim to capitalize on the advantages of different materials, leading to enhanced performance, bandwidth, and tunability.

3.1 THz horn antennas

These are specialized antennas designed to efficiently radiate and receive terahertz (THz) electromagnetic waves. THz horn antennas typically consist of a tapered metallic structure that transitions from a smaller aperture to a larger opening. This shape resembles a horn, hence the name. The inner surface of the horn is usually made of conductive material, which can be smooth or textured to optimize radiation efficiency. There are different types of horn antennas depending upon their shapes, such as:

Pyramidal Horn antenna: A common type that has a square or rectangular cross-section. It is widely used due to its good directivity and gain.

Conical Horn antenna: This type features a conical shape and is effective for applications requiring wide beamwidth (**Figure 6**).

Different advantages of THz horn antennas are to achieve significant gain, making them effective for applications requiring focused energy transmission and reception. These antennas exhibit good performance over a wide frequency range, making them suitable for various THz applications. The smooth inner surface reduces losses associated with reflections, improving efficiency. The relatively straightforward geometry allows for easier fabrication compared to more complex antenna designs. The high frequency of THz waves leads to smaller wavelengths, which results in compact antenna designs. However, this miniaturization presents several challenges, particularly regarding the fabrication and performance of THz horn antennas. The tip end of the horn, where the waves are emitted, requires very fine machining or fabrication techniques to maintain the correct shape and dimensions. Advanced techniques such as microfabrication or 3D printing may be required fr manufacturing these antennas.

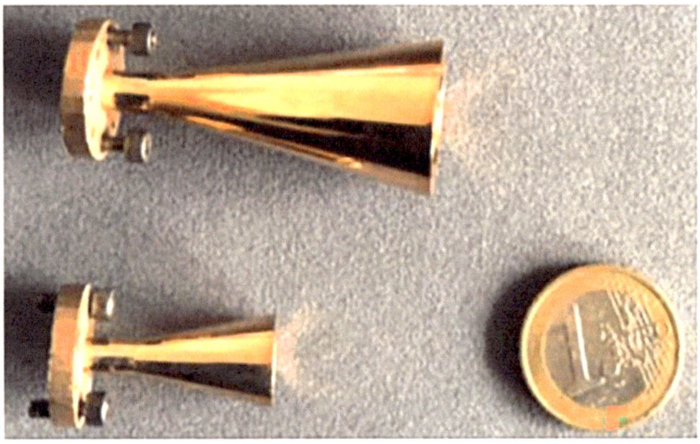

Figure 6.
Photograph of SLM-printed THz conical horn antennas [25].

3.2 THz dielectric antennas

Dielectric antennas are typically made from low-loss dielectric materials that can be shaped into various forms, such as patches, resonators, or waveguides. Common configurations include planar patch antennas and dielectric resonator antennas, which can be optimized for THz frequencies. Common dielectric materials include polymers (like PTFE and polyethylene), ceramics, and specially engineered composites. These materials are chosen for their low loss at THz frequencies and good thermal stability. Dielectric antennas can exhibit lower dielectric losses compared to metallic antennas, particularly in the THz range, which enhances overall efficiency. These antennas can be designed to operate over a wide frequency range, making them versatile for different THz applications. These antennas can often be made smaller than their metallic counterparts while still achieving effective radiation characteristics, aiding integration into compact systems. The dielectric constant can often be tuned by altering the material composition or structure, allowing for customization based on specific application needs. The thick medium mode occurs when the thickness of the dielectric substrate becomes comparable to the wavelength of the THz waves. This results in the excitation of surface waves that propagate along the interface of the dielectric materials. Surface waves are guided by the dielectric interface and can enhance the effective radiation and reception of THz signals. These waves are typically bound to the surface, leading to better energy confinement. The interaction between the dielectric substrate and the dielectric antenna can lead to better coupling of electromagnetic energy, enhancing overall efficiency. Surface waves can localize electromagnetic fields near the dielectric interface, increasing the antenna's sensitivity and effectiveness. As the frequency approaches the THz range, the behavior of the dielectric materials and the resulting surface wave phenomena become more pronounced. Surface waves can radiate energy away from the antenna, especially if the surface waves are not effectively confined. This can lead to reduced efficiency in the overall radiation pattern. The overall efficiency of the THz antenna can be compromised if a significant portion of the energy is lost to these mechanisms, making it less effective for applications requiring high sensitivity or power. To carefully balance the benefits of surface wave effects (such as enhanced coupling and sensitivity) with the potential for increased losses, sophisticated design and optimization are required. Loading lenses on THz dielectric antennas utilizes the antenna bunching property to effectively increase gain and improve overall performance. Lenses help refract and focus THz waves emitted from dielectric antennas, concentrating energy into a narrower beam. This focus can significantly enhance the antenna's effective gain. By shaping the phase fronts of the emitted or received waves, lenses enable constructive interference in the desired direction, leading to a more directive radiation pattern. The addition of a lens increases the electric field strength in the vicinity of the dielectric antenna, improving the overall efficiency of radiation [26]. It can help suppress unwanted side lobes in the radiation pattern, directing more energy into the main lobe and increasing gain (**Figure** 7).

3.3 Carbon nanotube dipole antennas

Carbon nanotube (CNT) dipole antennas are an intriguing topic in terahertz (THz) technology. These antennas leverage the unique electrical and mechanical properties of carbon nanotubes (CNTs) to operate efficiently at THz frequencies. In the case of CNTs, the nanotubes act as the arms of the dipole. The length of the CNTs typically corresponds to half the wavelength of the THz signal, which allows them to resonate effectively at lower THz frequencies. CNTs exhibit high electron mobility, which enhances their performance at high frequencies like THz. Their flexibility allows for novel applications in wearable or flexible electronics. CNT dipole antennas can be used in THz wireless communication systems for high-speed data transfer. They can serve as sensitive sensors in THz spectroscopy for detecting various materials. Producing CNT dipole antennas with precise dimensions and alignment can be challenging. Integrating these antennas with existing electronic systems requires innovative design solutions. The work by Samir F. Mahmoud and Ayed R. AlAjmi in 2012 introduced an innovative terahertz (THz) antenna structure utilizing carbon nanotubes (CNTs) wrapped in dielectric layers.

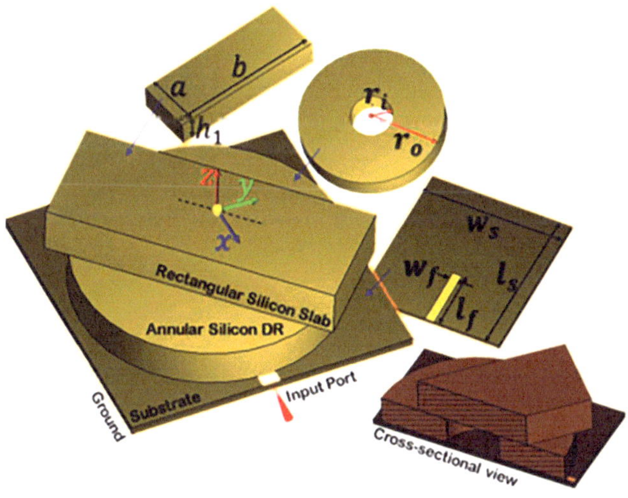

Figure 7.
Geometry of THz dielectric antenna [21].

3.4 Planar graphene THz antennas

These antennas typically have a two-dimensional structure, allowing for easy integration into various devices. Common designs include patch antennas, dipole antennas, and array configurations. The use of single or multiple layers of graphene can significantly affect the antenna's performance, including its resonance and bandwidth. Graphene's excellent electronic properties enable efficient operation at THz frequencies, allowing for rapid modulation and high-speed signal transmission. Graphene can support surface plasmons, enhancing the interaction of electromagnetic waves with the antenna, which can lead to improved performance in terms of gain and directivity. The resonance frequency of planar graphene antennas can be tuned by varying the gate voltage, enabling dynamic frequency adjustment. This tunability is particularly useful for applications requiring adaptability. While traditional antennas might be narrowband, graphene antennas can be designed to operate over a wider frequency range, which is beneficial for multi-frequency applications. Creating high-quality graphene layers and integrating them into antenna designs can be technically challenging. While graphene is a good conductor, it can still exhibit losses that may affect the efficiency of the antennas, especially at higher frequencies. Planar graphene THz antennas hold great promise for future technological advancements, offering unique properties that can enhance performance in communication, sensing, and imaging applications (**Figure 8**).

3.5 Terahertz photoconductive antennas

These antennas exploit photoconductivity to generate and detect terahertz (THz) radiation. These antennas have gained significant attention due to their potential in various applications, including imaging, communication, and spectroscopy. Terahertz photoconductive antennas (TPCAs) are typically made from semiconductors that exhibit increased electrical conductivity when illuminated by a laser. This process generates charge carriers (electrons and holes) in the material. When a laser pulse is directed onto the antenna, the generated charge carriers are accelerated by a bias voltage applied across the antenna structure. This acceleration produces THz

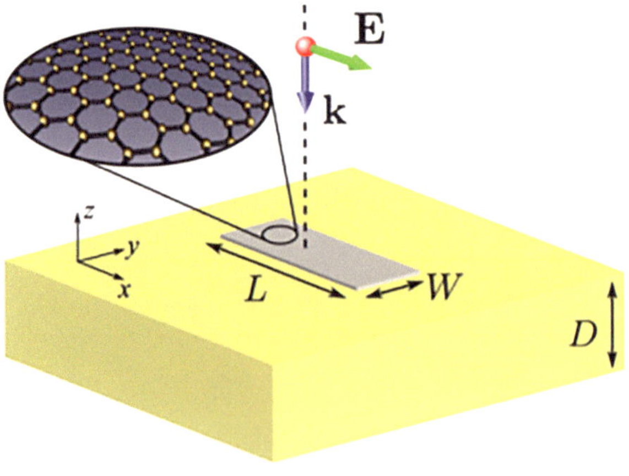

Figure 8.
Configuration of graphene-based THz nano-antenna [27].

electromagnetic pulses. These antennas can also be used for detecting THz radiation. Incoming THz pulses can modulate the conductivity of the antenna, which can then be measured as a voltage signal. TPCAs often feature a dipole or log-periodic antenna structure to efficiently couple to THz radiation. The design can be optimized for specific frequency ranges. Common materials for TPCAs include low-temperature grown gallium arsenide (LT-GaAs), indium antimonide (InSb), and other semiconductors known for their photoconductive properties. The choice of the laser wavelength is crucial, as it must match the absorption characteristics of the semiconductor to effectively generate charge carriers. These antennas can operate over a wide frequency range, making them suitable for various THz applications. Their ability to detect weak THz signals with high sensitivity is beneficial for spectroscopy and imaging. TPCAs can be fabricated in small sizes, allowing for integration into portable devices. The efficiency of TPCAs depends on the semiconductor material and its quality, which can affect performance. Precise fabrication techniques are required to create effective antenna designs and ensure optimal performance (**Figure 9**) [29].

3.6 THz lens antennas

These antennas are designed to enhance the performance of THz systems by focusing and directing THz waves. They combine the principles of lens optics with antenna technology to achieve improved radiation patterns, gain, and efficiency. THz lens antennas use materials with specific refractive indices to focus THz radiation, similar to how optical lenses focus visible light. This allows for more effective transmission and reception of THz signals. Common materials include low-loss

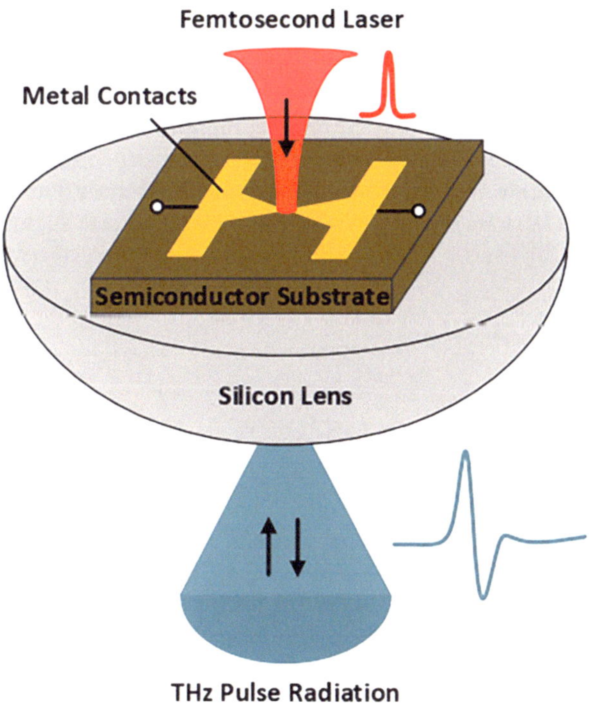

Figure 9.
THz photoconductive antenna [28].

dielectrics like polymers or crystals that are transparent at THz frequencies, such as Teflon, PTFE, or even certain types of ceramics. The design helps achieve high directivity, allowing the antenna to transmit or receive THz signals from specific directions, which is beneficial in communication systems. The shape of the lens is critical. Common designs include parabolic, hyperbolic, or Fresnel lenses tailored for THz wavelengths to optimize focus and minimize losses. The lens is often integrated with THz antennas (like dipole or patch antennas) to improve performance. This integration helps match the impedance and enhances the overall efficiency of the system. Advanced designs may incorporate multilayer structures or metamaterials to manipulate the phase of THz waves, providing additional control over the radiation pattern. THz lens antennas are used in imaging applications, including biomedical imaging, security scanning, and non-destructive testing, allowing for high-resolution imaging capabilities. They can enhance THz communication systems by focusing on signals, improving data rates, and extending the range of transmission. In THz spectroscopy, lens antennas can enhance signal collection efficiency, enabling detailed analysis of material properties. While lens materials can be selected for low loss, there are still challenges related to absorption and scattering at THz frequencies. Achieving the necessary precision in lens fabrication is critical for optimal performance, as any imperfections can degrade focusing capabilities. Developing high-performance THz lens antennas can be costly and complex, particularly when using advanced materials or fabrication techniques (**Figure 10**).

3.7 Terahertz microstrip antennas

It utilizes microstrip technology for efficient operation in the terahertz (THz) frequency range. They are characterized by their low profile, ease of fabrication, and integration capabilities with other electronic components. These antennas typically consist of a thin metallic patch (radiating element) printed on a dielectric substrate with a ground plane on the opposite side. The patch can be designed in various shapes, such as rectangular, circular, or more complex geometries, to optimize performance. Common dielectric materials for THz microstrip antennas include low-loss substrates like polyimide, Teflon, or specific ceramic materials that remain effective at THz frequencies. Various feeding methods can be employed, including microstrip lines, coaxial feeds, or aperture coupling, depending on the design requirements and

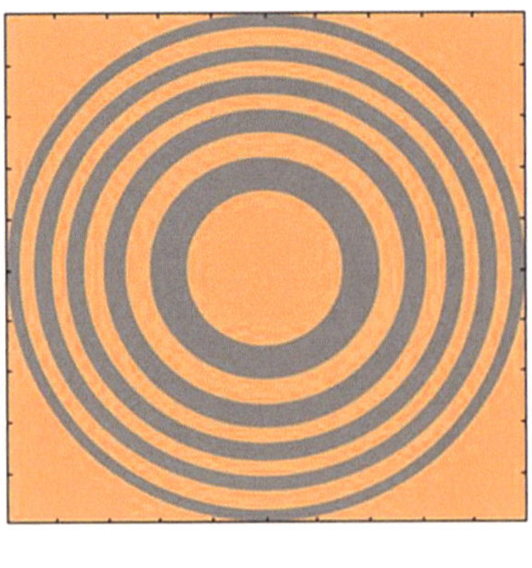

Figure 10.
Photograph of THz lens antenna [30].

desired impedance matching. Microstrip antennas can be designed to operate over a broad range of THz frequencies, making them suitable for multiple applications. Their thin, flat profile allows for integration into compact devices, which is particularly useful in portable applications. Microstrip antennas can be manufactured using standard PCB fabrication techniques, which simplifies production and reduces costs. The radiation patterns of microstrip antennas can be tailored by adjusting the shape and dimensions of the patch, allowing for directional or omnidirectional patterns as needed. While microstrip antennas typically have lower gain compared to larger antennas, careful design can enhance their efficiency and radiation characteristics, especially when combined with lens or reflector structures. These antennas are used in high-speed wireless communication systems, facilitating data transmission in the THz range. Microstrip antennas can be employed in THz imaging systems for applications such as security screening and biomedical imaging. The choice of substrate and metallic materials can affect performance due to losses, particularly at higher frequencies. Achieving good impedance matching across a wide frequency range can be challenging, requiring careful design. Microstrip antennas generally offer lower gain compared to other antenna types, which may necessitate the use of arrays or additional amplifying elements in some applications [31].

3.8 Terahertz on-chip antennas

These antennas are designed to be integrated directly onto semiconductor chips, enabling compact, efficient devices for THz signal generation and detection. On-chip antennas are typically fabricated alongside integrated circuits (ICs) using standard semiconductor processes, allowing for high levels of integration and miniaturization. Common designs include dipole antennas, patch antennas, and slot antennas. The choice of design depends on the desired radiation pattern, bandwidth, and integration requirements. Silicon, GaAs, and other semiconductor materials are often used as substrates due to their favorable electronic properties and compatibility with existing fabrication techniques. On-chip antennas are designed to be small, enabling them to fit within the limited space of integrated circuits while maintaining performance at THz frequencies. Many on-chip antennas can be designed for broadband operation, which is advantageous for THz applications that require flexibility across multiple frequencies. The choice of materials and careful design can minimize losses, enhancing the overall efficiency of THz signal transmission and reception. The design of on-chip antennas allows for the optimization of radiation patterns, enabling targeted signal transmission and reception, which is crucial for effective communication systems. Achieving good impedance matching is essential for efficient energy transfer, and careful design considerations are necessary to ensure compatibility with the connected ICs. On-chip antennas are pivotal in developing high-speed wireless communication systems operating in the THz frequency range, enabling data rates significantly higher than current technologies. They can be used in integrated THz sensors for applications such as environmental monitoring, security screening, and biomedical diagnostics. In THz imaging systems, on-chip antennas facilitate compact setups for non-destructive testing and medical imaging. The choice of semiconductor materials can introduce losses at THz frequencies, which may impact the efficiency of on-chip antennas. Achieving the required precision in the fabrication of small structures is critical for optimal antenna performance, and variations can lead to degraded characteristics. As with any integrated device, managing heat dissipation in THz systems is crucial, particularly in high-power applications.

4. Challenges of terahertz antennas

The development of terahertz (THz) antennas is becoming increasingly important due to the growing demand for spectrum resources in communication and sensing applications. With the rise of mobile devices, IoT applications, and high-speed data requirements, the demand for available spectrum is reaching critical levels, particularly in lower frequency bands. The development of THz antennas in response to the shortage of spectrum resources is a critical area of focus. By leveraging the underutilized THz spectrum, these antennas have the potential to support high-capacity communication and advanced sensing applications, helping to address the challenges of an increasingly crowded frequency landscape. The relationship between resonant frequency, antenna size, and losses is a critical consideration in the design of THz antennas. While higher frequencies allow for miniaturization, they also bring challenges related to increased losses and reduced efficiency. The physical dimensions of antennas are inversely proportional to frequency. Higher resonant frequencies mean shorter wavelengths, resulting in smaller antenna designs. For example, a THz antenna operating at 1 THz has a wavelength of about 0.3 mm, leading to very small antenna structures. At higher frequencies, currents tend to concentrate near the surface of conductors (skin effect), which increases the effective resistance and leads to higher resistive losses. The lack of perfectly matched substrate materials for terahertz (THz) antennas poses significant challenges in achieving optimal performance. Substrate materials that do not match the impedance of the antenna can lead to significant reflection losses, reducing the efficiency of signal transmission and reception. Inconsistent substrate properties can cause variability in performance across different antennas, making it difficult to standardize designs. Many commonly used dielectric materials exhibit high-loss tangents at THz frequencies, leading to significant energy dissipation as heat and reducing overall antenna efficiency. The dielectric properties of materials often change with frequency, complicating the design and making it challenging to find materials that perform well across the entire THz spectrum. The range of materials suitable for THz applications is relatively narrow. Traditional substrates used in microwave applications often do not perform well at THz frequencies. Advanced materials that show promise for THz applications can be expensive or challenging to fabricate, limiting their widespread use. Substrate materials must also maintain stability under varying thermal conditions. Poor thermal management can lead to performance degradation. Research into new low-loss dielectric materials, such as certain polymers or ceramics, may yield suitable substrates for THz antennas. These engineered materials can be designed to exhibit specific electromagnetic properties, potentially providing better matching for THz applications. Using layered structures with different materials can help create a more favorable environment for the antenna, balancing performance and loss characteristics. Combining different materials to optimize properties such as dielectric constant, loss tangent, and thermal stability can improve overall performance. Incorporating tunable elements that can adjust to changing conditions or frequencies can help mitigate the impact of substrate mismatches. Developing antenna designs that can compensate for substrate-related issues through innovative geometries or configurations. THz antennas often require high precision in their dimensions due to the small wavelengths involved. Achieving this level of precision can necessitate advanced, expensive fabrication techniques. Methods like photolithography or microfabrication are commonly used but can be costly and time-consuming. These techniques may not be easily scalable for mass production. The need for specialized materials with low dielectric losses suitable for THz applications can significantly increase costs. Many of these materials are not readily available or are expensive to process. Inconsistent material quality can lead to variations in antenna performance, requiring additional testing and quality control that can drive up costs. Manufacturing THz antennas often requires a high level of expertise, which can lead to increased labor costs. Skilled technicians are needed for both fabrication and testing. The need for specialized training to handle advanced manufacturing processes can add to overall expenses. Ensuring that THz antennas meet performance specifications requires extensive testing, which can be resource-intensive and costly. Each antenna may need individual calibration, adding further complexity and expense to the manufacturing process. Many THz antenna designs are still in the research phase, leading to low production volumes that do not benefit from economies of scale. Tailoring antennas for specific applications can complicate manufacturing processes and drive up costs.

5. Terahertz octagonal microstrip antenna design

A terahertz antenna is created using the CST simulation tool while rigorously following the aforementioned constraints. The operational frequency of an antenna

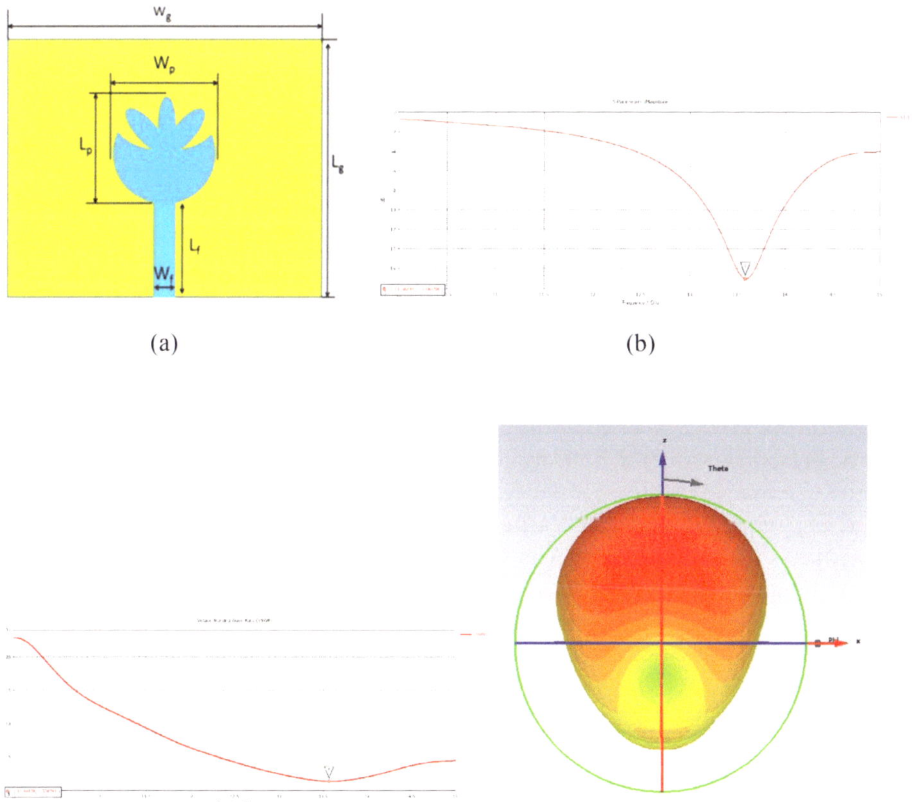

Figure 11.
(a) Proposed terahertz octagonal microstrip antenna configuration, (b) S-parameter response of the proposed antenna, (c) VSWR response of the proposed antenna, (d) farfield three-dimensional response of the proposed antenna.

is the frequency at which it receives and/or transmits signals. Whenever the patch height is specified, it may be estimated, or it can be selected well before design. The operating frequency of this design is 10 GHz. The height of the patch antenna can be set before identifying the operating frequency of the antenna, or the frequency band could be used to define the height, or both can be decided before the design. The proposed antenna is 1.0 mm tall. An antenna with such a full ground plane and no notch has been structured and is depicted in **Figure 11a**. The layout is printed with such a raised monopole on alumina material that has a dielectric constant of 4.3, permeability of one, loss tangent of 0.02, high mechanical properties, and insulating properties. The antenna's overall dimensions are 10.2 mm (height) × 4.8 mm (length) × 1.6 mm (width or substrate thickness). A feeding line and a large rectangular form with an inset microstrip.

6. Future work and conclusion

Future research directions for terahertz (THz) antennas are crucial for advancing their applications in communication, sensing, and imaging. Research into new materials with low dielectric losses that perform well at THz frequencies is essential. This includes exploring novel dielectrics, metamaterials, and nanomaterials. Investigating materials that can be easily integrated into flexible electronics or wearable devices can expand the applicability of THz antennas. Developing innovative designs that allow for smaller antennas without sacrificing performance, such as fractal or multilayer structures. Leveraging the unique properties of metamaterials and plasmonic structures to enhance gain, directivity, and bandwidth. Researching methods to integrate THz antennas with other components, such as photonic circuits and semiconductor devices, to create compact, multifunctional systems. Exploring the combination of THz antennas with technologies like terahertz time-domain spectroscopy (THz-TDS) for enhanced sensing capabilities. Focusing on methods to increase the efficiency of THz antennas, including impedance matching techniques and the use of advanced tuning mechanisms. Developing antennas that can operate effectively over a wider frequency range to support diverse applications. Investigating the use of artificial intelligence and machine learning for dynamic control and optimization of THz antenna performance in real-time applications. Researching advanced beamforming strategies to enhance the directionality and signal quality of THz communication systems. Exploring the use of THz antennas in non-invasive imaging and sensing for medical diagnostics, including cancer detection and tissue analysis. Developing THz antennas for applications in environmental sensing, such as detecting pollutants or analyzing atmospheric conditions. Investigating low-cost and scalable manufacturing techniques, such as additive manufacturing, to enable widespread production of THz antennas. Developing advanced testing and quality assurance methods to ensure consistent performance across mass-produced antennas. The future of terahertz antennas holds significant promise across various fields, but addressing the challenges and exploring innovative research directions is crucial. By focusing on materials, design, integration, and applications, researchers can pave the way for the successful deployment of THz antennas in communication, sensing, and beyond.

Terahertz (THz) antennas represent a rapidly evolving area of research and technology with the potential to revolutionize various fields, including communications, sensing, and imaging. As the demand for higher data rates and more sophisticated sensing applications grows, THz antennas offer unique advantages due to their ability to operate in the THz frequency range, which lies between microwave and infrared. THz antennas can facilitate ultra-high-speed wireless communication, enabling advancements in data transfer rates and bandwidth that traditional technologies cannot achieve. Ongoing research into low-loss materials, advanced design methodologies, and novel fabrication techniques is essential for enhancing the efficiency, performance, and miniaturization of THz antennas. The versatility of THz antennas extends to various applications, including biomedical sensing, environmental monitoring, and security screening, highlighting their potential impact on both industry and daily life. While the potential is significant, challenges remain in terms of manufacturing costs, scalability, and the availability of suitable materials. Addressing these challenges will be crucial for the commercial viability of THz technologies.

Author details

Somdotta Roy Choudhury
Department of Electronics and Communication Engineering, Ramrao Adik Institute of Technology, Nerul, Navi Mumbai, Maharashtra, India

*Address all correspondence to: somdottaroychoudhury@gmail.com

IntechOpen

© 2025 The Author(s). Licensee IntechOpen. This chapter is distributed under the terms of the Creative Commons Attribution License (http://creativecommons.org/licenses/by/4.0), which permits unrestricted use, distribution, and reproduction in any medium, provided the original work is properly cited.

References

[1] Krishna Ch M, Islam T, Suguna N, Kumari SV, Devi RDH, Das S. A micro-scaled graphene-based wideband (0.57-1.02 THz) patch antenna for terahertz applications. Results in Optics. 2023;**12**:1-18

[2] Rogalski A, Sizov F. Terahertz detectors and focal plane arrays. Opto-Electronics Review. 2011;**19**(3):346-404

[3] Malhotra I, Jha KR, Singh G. Terahertz antenna technology for imaging applications: A technical review. International Journal of Microwave and Wireless Technologies. 2018;**10**:1-20. DOI: 10.1017/S175907871800003X

[4] Xu F, Lin Y, Huang J, et al. Big data driven mobile traffic understanding and forecasting: A time series approach. IEEE Transactions on Services Computing. 2016;**9**(5):796-805

[5] Venkata Ratnam P, Sudhakiran B, Krishna B. A brief review on development of terahertz antennas. International Journal of Engineering Research and Applications. 2023;**13**(7):98-108

[6] Guan K, Li G, Kürner T, Molisch AF, et al. On millimeter wave and THz mobile radio channel for smart rail mobility. IEEE Transactions on Vehicular Technology. 2017;**66**(7):5658-5674

[7] Tabata H. Application of terahertz wave technology in the biomedical field. IEEE Transactions on Terahertz Science and Technology. 2015;**5**(6):1146-1153

[8] Nagatsuma T. Antenna technologies for terahertz communications. In: Proc. 2018 International Symposium on Antennas and Propagation (ISAP). Korea: IEEE; 2018. pp. 1-2

[9] Xu Z, Dong X, Bornemann J. Design of a reconfigurable MIMO system for THz communications based on graphene antennas. IEEE Transactions on Terahertz Science and Technology. 2014;**4**(5):609-617

[10] Rebeiz GM. Millimeter-wave and terahertz integrated circuit antennas. Proceedings of the IEEE. 1992;**80**(11):1748-1770

[11] Mahmoud SF, AlAjmi AR. Characteristics of a new carbon nanotube antenna structure with enhanced radiation in the sub-terahertz range. IEEE Transactions on Nanotechnology. 2012;**11**(3):640-646

[12] Chen S et al. Beam-space multiplexing: Practice, theory, and trends, from 4G TDLTE, 5G, to 6G and beyond. IEEE Wireless Communications. April 2020;**27**(2):162-172. DOI: 10.1109/MWC.001.1900307

[13] Nagatsuma T et al. Terahertz wireless communications based on photonics technologies. Optics Express. 2013;**21**(20):23736-23747

[14] Tsujimura K et al. A causal channel model for the terahertz band. IEEE Transactions on Terahertz Science and Technology. 2018;**8**(1):52-62

[15] Song H-J, Ajito K, Muramoto Y, Wakatsuki A, Nagatsuma T, Kukutsu N. 24 Gbit/s data transmission in 300 GHz band for future terahertz communications. Electronics Letters. 2012;**48**(15):953-954

[16] Takahashi H et al. 10-Gbit/s BPSK modulator and demodulator for a 120-GHzband wireless link. IEEE Transactions on Microwave Theory and Techniques. 2011;**59**(5):1361-1368

[17] Takahashi H et al. 120-GHz-band BPSK modulator and demodulator

for 10-Gbit/s data transmission. In: Proc. 2009 IEEE MTT-S International Microwave Symposium Digest. Boston, MA: IEEE; 2009. pp. 557-560

[18] IEEE standard for high data rate? Wireless multi-media networks--amendment 2:100 Gb/s wireless switched point-to-point physical layer. In: IEEE Std 802.15.3d-2017 (Amendment to IEEE Std 802.15.3-2016 as Amended by IEEE Std 802.15.3e-2017). SA: IEEE; 2017. pp. 1-55 White Paper

[19] Kosugi T, Tokumitsu M, Enoki T, Muraguchi M, Hirata A, Nagatsuma T. 120-GHz Tx/Rx chipset for IO-Gbit/s wireless applications using 0.1-μm-gate InP HEMTs. In: Proc. IEEE Compound Semiconductor Integrated Circuit Symposium. China: China Institute of Communications and IEEE Communications Society; 2004. pp. 171-174

[20] Hirata A, Kosugi T, Takahashi H, Yamaguchi R, Nakajima F, Furuta T, et al. 120-GHz-band millimeter-wave photonic wireless link for 10- Gb/s data transmission. IEEE Transactions on Microwave Theory and Techniques. 2006;**54**(5):1937-1944

[21] Gupta R, Varshney G, Yaduvanshi RS. Tunable terahertz circularly polarized dielectric resonator antenna. Optik – International Journal for Light and Electron Optics. 2021;**239**:1-7

[22] Yu X et al. Direct terahertz communications with wireless and fiber links. In: Proc. 2019 44th International Conference on Infrared, Millimeter, and Terahertz Waves (IRMMW-THz). Paris: IEEE; 2019. pp. 1-2

[23] Nagatsuma T et al. 12.5-Gbit/s wireless link at 720 GHz based on photonics. In: Proc. 2019 44th International Conference on Infrared, Millimeter, and Terahertz Waves (IRMMW-THz). Paris: IEEE; 2019. pp. 1-2

[24] Khang Nguyen T. Photoconductive dipole antennas for efficient terahertz receiver. Optics Communications. 2017;**383**:50-56

[25] Otter WJ. Hybrid 3-D-printing technology for tunable THz applications. Proceedings of the IEEE. 2017;**105**(4):756-767

[26] Withayachumnankul W, Fujita M, Nagatsuma T. Polarization responses of terahertz dielectric rod antenna arrays. In: Proc. 2019 International Conference on Microwave and Millimeter WaveTechnology (ICMMT). China: IEEE; 2019. pp. 1-3

[27] Llatser I et al. Graphene-based nano-patch antenna for terahertz radiation. Photonics and Nanostructures – Fundamentals and Applications. 2012;**10**:353-358

[28] Chen P, Hosseini M, Babakhani A. An integrated germanium-based THz impulse radiator with an optical waveguide coupled photoconductive switch in silicon. Micromachines. 2019;**10**:376-386

[29] Headland D, Fujita M, Nagatsuma T. Bragg-mirror suppression for enhanced bandwidth in terahertz photonic crystal waveguides. IEEE Journal of Selected Topics in Quantum Electronics. 2020;**26**(2):1-9

[30] Al-Dafaie S et al. Design and implementation of a terahertz lens-antenna for a photonic integrated circuits based THz systems. Springer Nature. 2022;**12**:1476

[31] Shihzad W et al. Design and analysis of dual-band high-gain THz antenna array for THz space applications. Applied Sciences. 2022;**12**(18):2-20

Section 3

THz Characterization

Chapter 4

Development and Application of a Terahertz Measuring System for Dielectric Characterization

Yiyang Luo, Olena Kryvenko, Vladyslav Lutsenko and Vitalii Yakovliev

Abstract

This study investigates the development and implementation of a system utilizing a gas laser for measuring dielectric characteristics within the terahertz frequency range. Methodological aspects include signal modulation using a shutter for enhanced radiometric reception and the integration of a reference channel to stabilize measurements against fluctuations in emitted power. Techniques for background compensation in both measurement and calibration channels are detailed, crucial for extending the dynamic range of attenuation measurements across diverse dielectric materials. Experimental results reveal linear attenuation coefficients for various materials, ranging from low values for polystyrene variants to significantly higher values for materials such as Getinax and organic glass. The findings underscore the feasibility of utilizing different dielectrics in terahertz applications based on their transparency coefficients.

Keywords: absorption characteristics, attenuation coefficient, terahertz waves, dielectric characterization, gas laser, radiometry, refractive index, submillimeter wavelengths

1. Introduction

Understanding the frequency behavior of complex permittivity at high frequencies is essential for elucidating the processes occurring during the interaction of matter with electromagnetic fields. The millimeter and submillimeter wavelength ranges, specifically spanning 10 to 0.1 mm (30 to 3000 GHz or 3×10^{10}–3×10^{12} Hz), are of particular interest due to the presence of dispersion regions for many substances within these frequencies. Consequently, the development of the terahertz (THz) frequency range (0.1–10 THz) represents one of the most dynamic and rapidly advancing areas in contemporary applied physics.

The terahertz (THz) range includes the rotational spectra of many organic molecules, collective modes in DNA and proteins, and phonon resonances in crystal lattices, enabling the development of advanced spectroscopic methods for biological and semiconductor materials [1]. THz waves, which can be reflected, absorbed, or transmitted by a medium, provide extensive information about the material being studied. This versatility supports a wide range of applications, such as controlling chemical reactions and manipulating electronic states in quantum wells [2, 3]. Unlike X-rays, THz radiation is non-ionizing and does not damage biomolecules, making it suitable for medical diagnostics, including cancer detection and assessing burns [4].

Measuring the dielectric properties of materials is crucial for the development and fabrication of microelectronic components. Advances in submicron technology have pushed component operational frequencies into the terahertz range. However, spectroscopy in this range poses significant challenges, as it extends beyond the effective limits of traditional microwave and optical methods. In the microwave range, where wavelengths are larger than the sample's characteristic dimensions, dielectric constants are commonly determined by measuring the capacitance of a resonator containing the sample [5, 6]. This technique allows for dielectric constant measurements across a broad range (1 to 300 Hz) up to 20 GHz. As the frequency increases, parameter extraction becomes more complex, and the requirements for sample surface quality become more stringent.

In the optical frequency range, Fourier-transform spectroscopy is the primary method for determining the dielectric constant with high precision [7, 8]. However, its accuracy diminishes with increasing wavelength, rendering it unsuitable for the submillimeter range. Currently, time-domain spectroscopy is widely utilized for many spectral measurements in the terahertz frequency range [9]. This technique demands high-quality sample surfaces and can exhibit resonance features related to sample geometry. Additionally, highly sensitive electromagnetic radiation receivers often operate at cryogenic temperatures (77 K and below), complicating traditional dielectric constant measurements at these low temperatures.

Josephson spectroscopy presents a promising alternative or supplementary measurement technique [10]. Its capabilities have been demonstrated for studying various sources of electromagnetic radiation in the terahertz range and for characterizing electrodynamic systems, such as planar periodic antennas (60 to 1.2 THz) and planar square resonators (50 to 700 GHz) [11].

One of the most challenging aspects of advancing terahertz technology is the development of efficient methods for generating and receiving coherent terahertz radiation. This frequency range, situated between the infrared and microwave regions, lacks well-established physical principles for radiation generation that have been developed over the past half-century for optical and microwave domains [12–15]. Conventional approaches from vacuum and solid-state electronics, as well as standard quantum electronics circuits, are ineffective in this range. Generating submillimeter backward wave oscillators (BWO) and optical parametric oscillators (OPO) faces issues such as low cathode efficiency, complex electron beam coordination with the delay system, and high losses in vacuum waveguides [14]. Molecular beam masers require substantial cooling and high-Q microresonators, while gas discharge lasers achieve sufficient inversion at rotational transitions for generation in only a limited number of lines [15]. Avalanche transit-time diodes and Gunn diodes are also ineffective at these wavelengths [12]. Exciting terahertz oscillations using short (femtosecond) optical pulses, which generate subpicosecond polarization bursts in cold plasmas or dipole photoconductive antennas, is limited by low process efficiency [16–18]. Additionally, current available sources of coherent terahertz radiation have low-power (nanowatts to microwatts, occasionally up to watts), are poorly tunable, and cover only narrow frequency bands.

For detecting terahertz radiation, broadband detectors such as bolometers [19–21], pyroelectric sensors, Schottky diodes (based on metal-semiconductor junction non-linearities) [22, 23], and photoconductive detectors [24] are predominantly used. The employment of selective and tunable detectors in spectral analysis can eliminate the need for diffraction gratings or mechanically tunable interferometers. The complexities in generating and receiving terahertz waves have resulted in relatively few experimental studies on the properties of materials and media in the submillimeter range [25–31].

This chapter is dedicated to the construction of measurement system, calibration methods, and the validation of results for measuring dielectric properties in the submillimeter range. This comprehensive approach will address current limitations and contribute to the advancement of terahertz spectroscopy and its applications.

2. Theoretical basis for THz wave interaction with dielectric materials

The characterization of dielectric materials in the terahertz (THz) frequency range is critical for various applications, including imaging, sensing, and communication. This chapter builds upon existing methodologies by implementing a gas laser-based system and providing an extensive theoretical analysis, including the comparison of different models to estimate dielectric properties accurately.

2.1 Gas laser-based THz radiation source

Gas lasers, such as the one used in your experimental setup, operate based on the principles of stimulated emission. The active medium, consisting of gases such as methane (CH_4) and nitrogen (N_2), is excited by an external energy source (in our research, a 12.3 MHz signal) [32]. This excitation leads to the population inversion necessary for lasing. The generation frequency is determined by the population level differences during laser excitation and the resonance frequencies of the Fabry-Perot resonator. The resonator's frequency (ν) response will display a linear pattern, with evenly spaced resonance lines:

$$\nu = \frac{c}{2L}\left(1 - \frac{\Delta n(p)}{n_0}\right), \tag{1}$$

where c is the speed of light in vacuum (3×10^8 m/s), L is the resonator length, $\Delta n(p)$ is the change in refractive index due to pressure p, and n_0 is the refractive index at standard pressure. The resulting variation in frequency (ν) affects the laser's output wavelength, ranging from 0.30 to 0.38 mm (790 to 1000 GHz). The frequency variation $\Delta\nu$ due to a small change in power Δp can be approximated using the following:

$$\Delta\nu \approx \left(\frac{\partial\nu}{\partial n} \times \frac{\partial n}{\partial p}\right)\Delta p. \tag{2}$$

In practical experiments, the frequency stability of terahertz lasers is influenced by several factors. Changes in the refractive index due to pressure variations directly affect the laser frequency, with higher pressure generally increasing the refractive index and potentially altering the frequency. Additionally, variations in power can cause thermal effects and nonlinear responses in the laser material, further impacting frequency stability. Although quantum effects are usually minor in standard applications, they may become significant in high-precision scenarios. Therefore, when designing and operating terahertz lasers, it is essential to consider these factors—pressure, power, and quantum effects—to ensure stable and accurate performance for applications such as imaging, material analysis, and communication.

2.2 Submillimeter wave generation, absorption, and detection

The generation and detection of submillimeter waves (or terahertz waves) involves complex interactions between electromagnetic fields and matter. In our research, the laser wavelength of 0.313 mm places this radiation in the terahertz range, which is particularly challenging due to its position between microwave and infrared frequencies. The radiation power, which fluctuates between 10 and 20 mW depending on the resonator length, is crucial for determining the intensity of the THz beam. Assuming a beam with a Gaussian profile, the intensity at a distance r from the beam center is given by following:

$$\boldsymbol{I}(r) = I_0 \exp\left(-\frac{2r^2}{\omega_0{}^2}\right), \tag{3}$$

where $I_0 = 2P/(\pi\omega_0{}^2)$ is the peak intensity, r is the radial distance from the beam axis in the plane perpendicular to the direction of propagation, P is the total power, and ω_0 is the beam waist. In our research, the thickness of material $\sim$0.1–50 mm. Assume $\omega_0 = 1$ mm, $I_0 \approx 6366–12732$ W/m^2, then at a distance of $r = 1$ mm from the beam center, we have $I(1 \text{ mm}) \approx 858–1716$ W/m^2. That means the intensity decreases exponentially with the square of the distance from the center of the beam, and at 1 mm from the center, the intensity drops to about 13.5% of its peak value.

For practical applications, it is often necessary to estimate the attenuation of THz waves in a material. If the thickness of the material is d, the transmitted intensity I_t can be related to the incident intensity I_0 through the Beer-Lambert law:

$$I_t = I_0 \exp(-\boldsymbol{\alpha d}). \tag{4}$$

Given that α is frequency-dependent, this relation allows us to calculate the material's transparency at specific THz frequencies. For materials with low α values, such as certain polymers, the transmission is high, making them suitable for THz applications. For instance, polystyrene variants typically exhibit low attenuation, while materials such as Getinax and organic glass have higher attenuation, which can be precisely quantified using this approach.

Let us make some specific estimation. The α for biological tissues typically range from 0.1 to 1 cm^{-1} [1]. For our experimental setup, let us assume we are working with a biological tissue or similar material where $\alpha = 0.3$ cm^{-1} and $d = 10$ mm, then we have $I_t \approx 74.08\%\ I_0$. Considering, $I_0 \approx 6366–12732$ W/m^2, $I_t \approx 4713.4–9426.4$ W/m^2.

2.3 Complex permittivity and complex refractive index

The interaction of THz radiation with dielectric materials is governed by the material's complex permittivity, $\varepsilon^* \equiv \varepsilon'\text{-}j\varepsilon''$, where ε' represents the material's ability to store electrical energy, and ε'' represents the dissipative losses. The propagation of

THz waves through a dielectric medium can be described by the complex wave vector **k**:

$$\mathbf{k} = \frac{\omega}{c}\sqrt{\boldsymbol{\varepsilon}^*} = \frac{\omega}{c}\sqrt{\boldsymbol{\varepsilon}' - j\boldsymbol{\varepsilon}''} = \boldsymbol{\beta} - j\boldsymbol{\alpha}. \tag{5}$$

The absorption and reflection of THz radiation are directly related to these parameters. The absorption coefficient α and the phase constant β are given by following:

$$\boldsymbol{\alpha} = \frac{\omega}{c}\mathrm{Im}\left(\sqrt{\boldsymbol{\varepsilon}^*}\right), \tag{6}$$

$$\boldsymbol{\beta} = \frac{\omega}{c}\,\mathrm{Re}\left(\sqrt{\boldsymbol{\varepsilon}^*}\right), \tag{7}$$

where ω is the angular frequency. These parameters determine how THz waves propagate through a dielectric material, with lower α values indicating higher transparency. The complex permittivity ε^* and the complex refractive index n^* are related through the equation.

$$\boldsymbol{\varepsilon}^* = (n^*)^2 = (n' - jn'')^2 = (\boldsymbol{n}')^2 - (n'')^2 - 2jn'n'', \tag{8}$$

where n' and n'' are real and imaginary parts of the refractive index, respectively. Thus.

$$\boldsymbol{\varepsilon}' = (\boldsymbol{n}')^2 - (n'')^2, \tag{9}$$

$$\boldsymbol{\varepsilon}'' = -2n'n''. \tag{10}$$

2.4 Transmittance coefficient of a plane-parallel plate

The transmittance T of a plane-parallel plate can be analyzed to determine material properties such as the absorption coefficient α, the real part of the refractive index n', and the imaginary part n'' [9]. In real experiments, it is essential to account for both the reflection of electromagnetic waves (EMWs) at multiple boundaries and their absorption within the material [33]. The thickness d of the plate can vary, being either small or large relative to the inverse of the absorption coefficient $1/\alpha$, which represents the distance over which the intensity decreases by a factor of e.

2.4.1 For strong absorption

For strong absorption, that is, αd is sufficiently large, allowing us to neglect multiple reflections within the plate, the transmittance T at normal incidence can be expressed as follows

$$T = \frac{|E_t|^2}{|E_i|^2} = (1 - R)^2 \exp(-\boldsymbol{\alpha} d). \tag{11}$$

where E_t and E_i are the transmitted and incident electric fields, respectively. Based on the formula above, the absorption coefficient α can be expressed as follows:

$$\boldsymbol{\alpha} = \frac{1}{d}\ln\left[\frac{(1 - R)^2}{T}\right] = -\frac{1}{d}\ln\left[\frac{T}{T_0}\right], \tag{12}$$

where T_0 represents the transmittance of the plate without absorption, serving as a reference value.

2.4.2 For weak absorption

For weak absorption, where $ad \ll 1$, the effects of multiple internal reflections within the plate are minimal. Under these conditions, the power reflection coefficient at the plate boundaries can be expressed as follows:

$$R = \frac{(n' - 1)^2 + n''^2}{(n' + 1)^2 + n''^2}, \tag{13}$$

where n' and n'' denote the real and imaginary parts of the refractive index, respectively. When the absorption is weak, specifically when $n''^2 << (n'-1)^2$, the reflection coefficient simplifies to:

$$R = \frac{(n' - 1)^2}{(n' + 1)^2}. \tag{14}$$

From this, we can solve for the real part of the refractive index n':

$$n' = \frac{1 + \sqrt{R}}{1 - \sqrt{R}}. \tag{15}$$

Since the EMWs undergoes reflection twice from the boundaries and is absorbed inside the plate, the coefficient determining the attenuation can be expressed as follows:

$$\alpha = \frac{2\pi n''}{\lambda} = kn'', \tag{16}$$

where λ is the wavelength of the electromagnetic wave in a vacuum and k is the wavenumber.

Furthermore, in scenarios where the absorption of EMWs by the material is minimal ($\alpha d \ll 1$), it is essential to account for multiple reflections within the sample plate. In such cases, the electromagnetic waves undergo several internal reflections between the surfaces of the sample, which can influence the overall measurement of transparency and absorption. Additionally, two extreme cases must be distinguished:

Non-interference case: Interference effects are negligible due to insufficient monochromaticity or plane-parallelism. In this case, phase relationships are not taken into account when summing the field strengths of the reflected beams, and only their intensities are added. The transmittance T can be expressed as follows:

$$T = \frac{(1 - R)^2 \left(1 + \frac{n''^2}{n'^2}\right) \exp(-\alpha d)}{1 - R^2 \exp(-2\alpha d)}. \tag{17}$$

In low absorption scenarios, where $\alpha d \ll 1$ and $n''^2/n'^2 \ll 1$, the transmittance T can be approximated as follows:

$$T \cong \frac{1 - R}{1 + R}. \tag{18}$$

In the case where $R^2\exp(-2\alpha d) < < 1$, the transmittance T can be approximated as follows:

$$T = \left[(1-R)^2\left(1+\frac{n''2}{n\prime^2}\right)\exp(-\alpha d)\right] \times \left[1 - R^2\exp(-2\alpha d)\right]. \tag{19}$$

This can be further expanded to:

$$\begin{aligned} T = &\left[(1-R)^2\exp(-\alpha d)\right] + \left[(1-R)^2\left(1+\frac{n''2}{n\prime^2}\right)\exp(-\alpha d)\right], \\ &+\left[(1-R)^2R^2\left(1+\frac{n''2}{n\prime^2}\right)\exp(-3\alpha d)\right] \approx T_0\left(1+\frac{n''2}{n\prime^2}\right) \end{aligned} \tag{20}$$

where T_0 is the transparency coefficient without background compensation. Solving for n'', we get

$$\boldsymbol{n}'' = n'\left(\sqrt{\frac{T}{T_0}} - 1\right). \tag{21}$$

Interference case: The radiation is monochromatic, and the plate is plane-parallel. It is necessary to consider interference and add the electric field strengths, considering phase losses φ at the boundaries and phase shift $\psi = 2\pi d{\cdot}n'/\lambda$ when the wave passes through the plate. In this scenario, the transmittance T is given by:

$$T = \frac{(1-R)^2\left[1+\frac{n''2}{n\prime^2}\exp(-\alpha d)\right]}{1 - 2R\exp(-\alpha d)\left[1 - 2\sin^2\left(\frac{2\pi dn'}{\lambda}+\varphi\right)\right] + R^2\exp(-2\alpha d)}. \tag{22}$$

Typically, in semiconductors, the refractive index is quite high ($n' = 3 \ldots 6$). If the absorption coefficient α is not very large ($\lesssim 10^3\ \text{cm}^{-1}$), then the ratio n''^2/n'^2 is much less than unity ($\ll 1$) and can be neglected in the numerator of Eqs. (19) and (22).

From Eq. (22), it follows that the transmittance T of monochromatic radiation with weak absorption in the plate ($\alpha d \ll 1$) should exhibit a periodic dependence on the wavelength with maxima at values of λ that satisfy the Fabry-Perot resonator condition:

$$2dn' \cong m\lambda, \tag{23}$$

where the integer m denotes the interference order. Minima T must occur under the condition:

$$2dn' \cong \left(m + \frac{1}{2}\right)\lambda. \tag{24}$$

In Eq. (23), an integer number of wavelengths fits on the doubled optical path $2dn'$; in Eq. (24), half-wavelengths fit. The approximate equality sign is used in Eq. (23) because a rigorous analysis requires consideration of phase losses ϕ, exact values of $\alpha{\cdot}d$, and their dispersion. To avoid dealing with half-integer numbers in Eq. (23), sometimes numbers m and $m + 1/2$ are replaced with twice as large numbers. In this case, even integers correspond to transmission maxima, and odd integers correspond to minima of interference transmission.

In the region of strong absorption ($\alpha d \gg 1$), the transmittance T is less than 0.1, and interference extremes are practically absent. In the transparency region ($\alpha{\cdot}d \ll 1$), clear maxima and minima are observed. Measuring the distance between the nearest maxima allows us to determine the interference order m by the following equations:

$$\Delta\lambda = \lambda_{m+1} - \lambda_m; \tag{25}$$

$$\Delta n' = n'_{m+1} - n'_m; \tag{26}$$

$$m = \frac{\lambda_{m+1}}{\Delta\lambda} + 2dn'\frac{\Delta \boldsymbol{n}'}{\Delta\lambda} \approx \frac{1}{\lambda_m}\left(\frac{1}{\lambda_{m+1}} - \frac{1}{\lambda_m}\right)^{-1}. \tag{27}$$

It should be noted that the dispersion of the refractive index $\Delta n'/\Delta\lambda$ is small in the transparency region, where the approximate equality in Eq. (27) holds true and increases near the region of strong absorption.

With weak absorption in the plate ($\alpha d \ll 1$), Eq. (22) suggests the feasibility of determining the refractive index when the dispersion of the refractive index n' is negligible. If (the exponential factors are close to unity) $n''^2/n'^2 \ll 1$, then from Eq. (13) and $\alpha d \ll 1$, we can determine the ratio of the transmittance coefficients at the minimum and maximum as follows:

$$\frac{T_{\min}}{T_{\max}} \approx \left(\frac{1-R}{1+R}\right)^2 = \left(\frac{2n'}{n'^2+1}\right)^2. \tag{28}$$

Solving for n', we get

$$n' \cong \sqrt{\frac{T_{\max}}{T_{\min}}} + \sqrt{\frac{T_{\max}}{T_{\min}} - 1}. \tag{29}$$

This approach assumes negligible absorption and dispersion effects. However, if absorption is significant ($\alpha d \gg 1$) or dispersion becomes non-negligible, the accuracy of this method decreases [33]. In strong absorption scenarios, interference patterns diminish, making transmittance measurements less reliable for determining n'. Hence, while this method is suitable for low absorption, it requires adjustments or alternative approaches for cases with high absorption or significant dispersion.

Having determined n' from Eq. (27) for the two nearest extrema, we can use these formulas to determine the thickness of the plate d. Since the value of d must be consistent across all extrema, calculations can be performed for multiple pairs of minima and maxima.

Once the value of d is known, Eq. (23) can then be employed to determine not only the value of n, but also its wavelength dependence λ. This procedure is well justified under conditions where both absorption and scattering of electromagnetic waves are minimal, and with high-quality surface treatment. Deviations from these conditions may result in overestimated $T_{\min}$ values, consequently leading to underestimated values of n^*.

3. Experimental setup

Measurement techniques commonly used for centimeter-wave frequencies face significant challenges at millimeter and submillimeter wavelengths due to the small size and precise calibration required. These difficulties are exacerbated by material processing limitations, affecting measurement accuracy. Optical methods, adapted from traditional optics, offer a viable alternative by focusing on electromagnetic waves into a narrow beam. However, practical implementation is complex, particularly in generating and detecting submillimeter waves. Our research contributes to addressing these challenges, and this section details the specific equipment and methodologies employed in our experimental setup.

3.1 Diagram and configuration of the experimental setup

In our research, the experimental setup for dielectric characterization in the terahertz range utilizes a two-channel system with a gas laser, as shown in **Figure 1**. The gas laser, detailed in Section 2, uses a methane-nitrogen mixture as its active medium to generate waves at approximately 960 GHz (0.313 mm wavelength), with the frequency varying from 790 GHz (0.38 mm) to 1000 GHz (0.30 mm) depending on pressure. The active medium is excited by a 12.3 MHz signal, producing radiation power between 10 and 20 mW as the resonator length is adjusted. Traditionally, gas lasers use direct current (DC) excitation to induce a discharge in the gas cavity and initiate oscillations within a Fabry-Perot resonator. However, DC excitation often results in soot accumulation on the electrodes, which deteriorates performance over time [33, 34]. Recent research indicates that alternating current (AC) excitation can reduce soot buildup and extend the laser's operational lifespan [35, 36].

The translucent/semi-transparent mirror is designed as a diffraction grating composed of thin (50 μm) conductors, allowing part of the signal to exit the resonator. Signal modulation is achieved using an obturator—a metal plate with 8 holes—driven by a synchronous motor at 50 Hz, resulting in a modulation frequency of 400 Hz. The modulated beam is divided into measurement and reference channels. **Figure 1** illustrates the experimental setup, while **Figure 2** presents two photos of the actual experimental scene. The numbered labels in the photos correspond to the modules shown in **Figure 1**.

The measurement setup involved plane-parallel plates placed along the laser signal path to the receiver. By recording the transmitted signal, the material's transparency was assessed. To address errors from fluctuating laser power, signal recordings were made in both the measurement system and the reference channel that monitored the laser power.

It is worth mentioning that for THz research there are challenges exist in receiving and generating transmitted signals in the THz range [33–36]. Modulation is essential for enhancing receiver sensitivity and enables the use of pyroelectric receiving

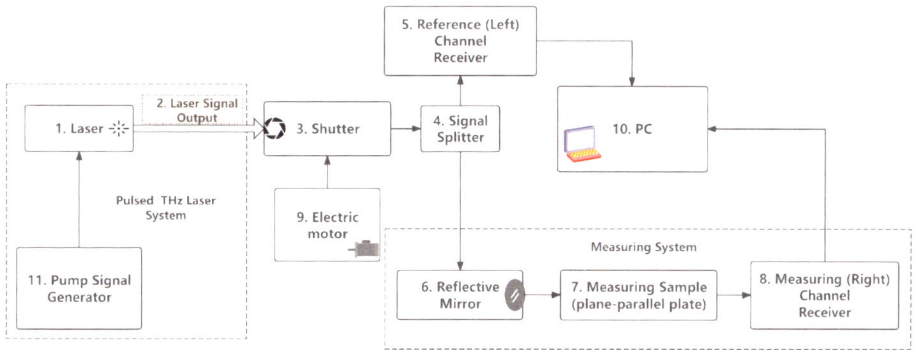

Figure 1.
A schematic diagram of the 2-channel measuring system for dielectric characterization using a terahertz gas laser system.

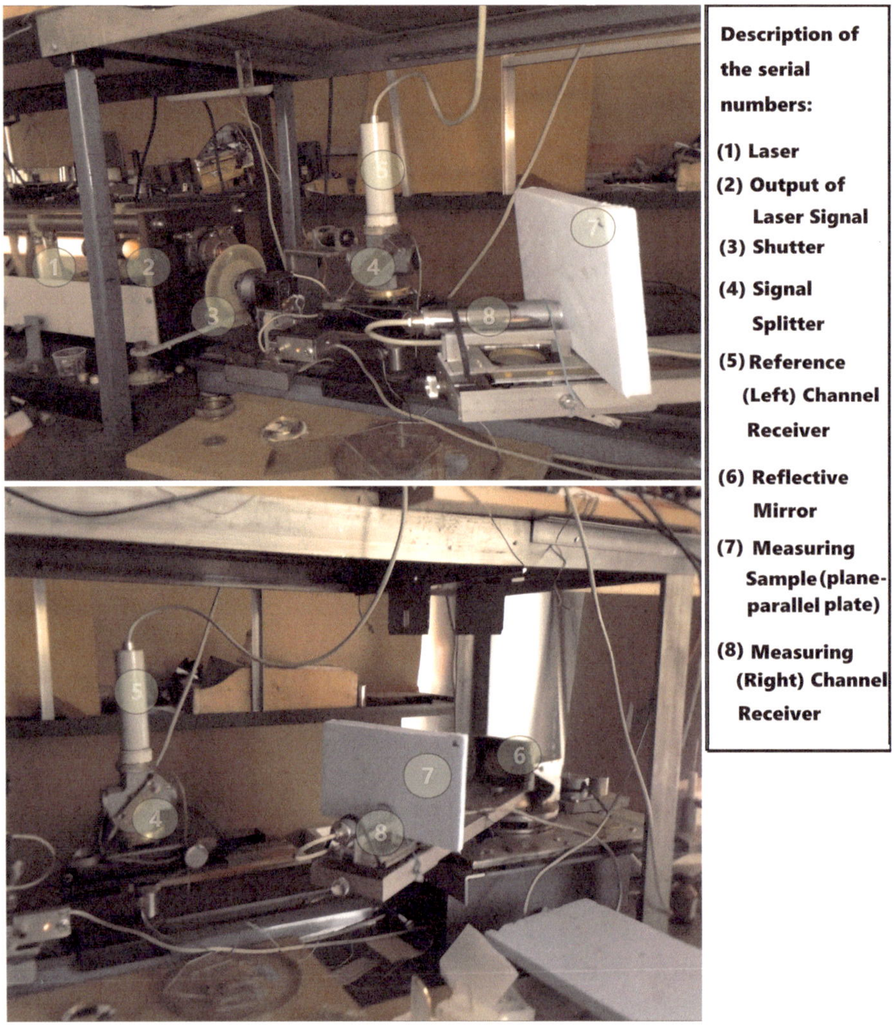

Figure 2.
Photograph of the experimental setup for dielectric property measurements using a terahertz gas laser system.

elements that do not respond to constant signals. Modulation and demodulation by reducing frequency offer several advantages:

1. *Stability and calibration for weak signals:* Similar techniques are widely used in radiometers to assess signal levels, ensuring that detected signals indicate normal operation.

2. *Dynamic range extension:* Modulation allows the receiving system to operate over a broader dynamic range, improving sensitivity for weak signals.

3. *Phase-locked detection:* Modulation facilitates narrowband filtering, enhancing the detection of weak signals.

Therefore, it is clear that the amplitude of the demodulated signal can be interpreted as representing the overall amplitude of the received terahertz signal. This demodulated amplitude accurately reflects the signal strength after the modulation and filtering processes, enabling a clearer analysis of the terahertz signal characteristics. A more detailed explanation will be provided in the next subsection.

3.2 Modulation technique and spectral analysis

To enhance narrow-band detection and measurement sensitivity, the setup employs signal modulation. The motor operates at 50 Hz, creating a laser signal modulation frequency of 400 Hz. As depicted in **Figures 1** and **2**, the disk with 8 holes enables narrowband isolation of the modulated signal through spectral analysis or fast Fourier transform (FFT), thereby improving measurement accuracy. In other words, this modulation technique brings "spectral coloring" to the signal and enhances narrowband reception at the modulation frequency, boosting the measurement system's effectiveness. The modulated beam is then directed through a signal separator, which divides it into a measurement channel and a calibration channel, as shown in **Figures 1** and 2.

This modulation can be represented mathematically as follows:

$$S(t) = A\cos(2\pi f_0 t)\cos(2\pi f_m t), \tag{30}$$

where A is the amplitude, f_0 is the frequency of the carrier (THz wave), and f_m is the modulation frequency. Applying trigonometric identities, the modulated signal can be decomposed into the following:

$$S(t) = \frac{A}{2}\left[\cos(2\pi(f_0 + f_m)t) + \cos(2\pi(f_0 - f_m)t)\right], \tag{31}$$

This decomposition indicates that the signal now contains two sidebands at frequencies $f_0 \pm f_m$, effectively shifting the signal away from low-frequency noise and enabling narrow-band detection.

To quantify the amplitude of signals in both the measurement and reference channels at the modulation frequency, an advanced algorithm was developed for automatic detection and measurement. This algorithm performs a FFT on a data segment consisting of $N = 2^{14} = 65{,}536$ samples, sampled at a rate of $f_s = 10$ kHz. This setup provides a frequency resolution $\Delta f = f_s/N = 0.153$ Hz and the Nyquist frequence $f_N = f_s/2 = 5$ kHz. The carrier frequency f_0 is well above this Nyquist limit, so direct detection of f_0 or its immediate sidebands $f_0 \pm f_m$ is not feasible. However, the modulation frequency f_m is much lower than the Nyquist frequency, that is, within the sampling range, and thus can be detected directly. To process the data, our algorithm applies a narrowband filter centered on 400 ± 10 Hz. This filter isolates the highest amplitude spectral component, ensuring accurate detection of the modulated signal despite background noise. From a signal processing perspective, the demodulation process at approximately 400 Hz is considered to be performed by this algorithm. **Figure 3** displays the spectra from both the reference and measurement channels across various experimental time intervals.

Additionally, the impact of engine rotational frequency instability on the modulation signal's spectral width was evaluated. **Figure 3** also demonstrates the signal stability and spectral characteristics, validating the effectiveness of the modulation technique and the measurement system's robustness under varying operating conditions. Allan variance analysis indicates that frequency deviations due to rotational instability are below 0.12 Hz.

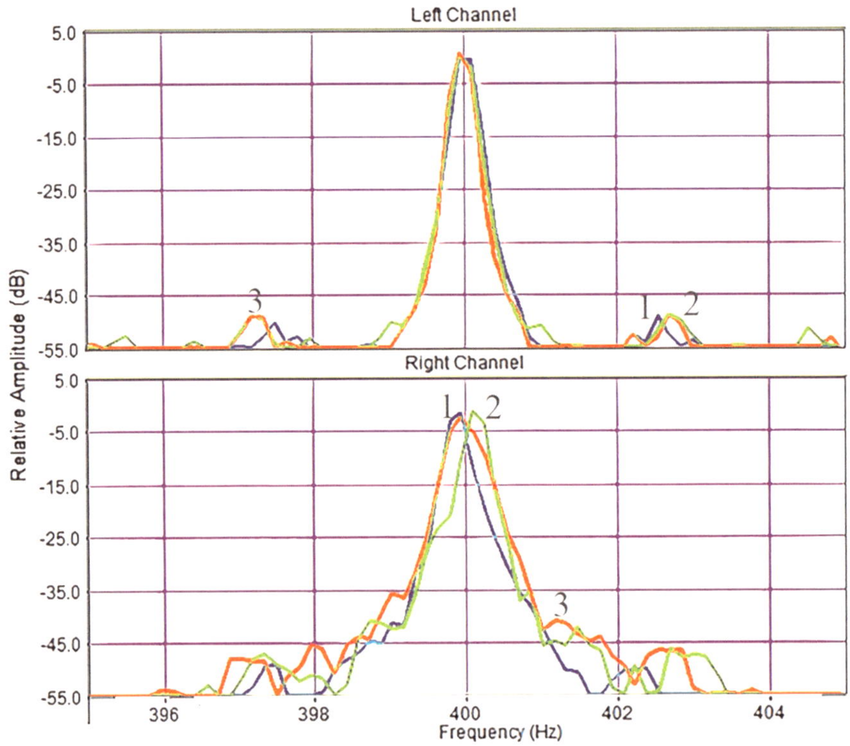

Figure 3.
Spectral analysis of signals from the reference (top) and measurement (bottom) channels across multiple experimental time intervals.

3.3 Algorithms for estimating material parameters

The experimental setup measures material transparency at terahertz frequencies using a gas laser (Module 1, as shown in **Figures 1** and 2). The laser signal, amplitude-modulated with an obturator (Module 3, as shown in **Figures 1** and 2) is shifted to a higher frequency to reduce low-frequency noise and enhance the signal-to-noise ratio (SNR), allowing precise material property estimation even with lower sampling rates. The setup includes two channels:

- *Measuring channel (module 5, see* **Figures 1** and 2*):* Contains the sample to measure the laser signal transmission through the material, denoted by $S_m(t)$.
- *Reference channel (modules 6, 7, 8, see* **Figures 1** and 2*):* Monitors fluctuations in the laser's power to account for output changes during the measurement, denoted $S_r(t)$.

Algorithm for measuring material transparency at terahertz frequencies by our experimental system can be described by the following steps:

- *Step 1: Initial calibration.* Measure and record the baseline signals from both the measuring channel ($S_{m(1)}(t)$) and the reference channel ($S_{r(1)}(t)$) without the sample. These values establish the reference points for subsequent measurements.

- *Step 2: Background noise compensation.* With a signal absorber placed at the laser output, measure and record the background noise signals in both the channels ($S_{m(2)}(t)$, $S_{r(2)}(t)$) to compensate for non-signal-related noise during analysis.

- *Step 3: Measurement with sample.* Insert the plate sample into the measuring channel and record the transmitted signals ($S_{m(k)}(t)$) and reference signals ($S_{r(k)}(t)$) for $k \geq 3$.

- *Step 4. Signal spectrum analysis and bandwidth filtering.* In the actual experiment, with a signal duration of 21 to 34 seconds and a sampling rate of 10 kHz, each segment comprises 65,536 samples. Consequently, three to five segments are available for analysis. Perform FFT to obtain the spectra of signal segments (see **Figure 4 (a)** and **(c)**). Apply bandpass filtering to isolate frequencies around 400 ± 10 Hz (see **Figure 4 (b)** and **(d)**). It is worth mentioning that the use of an absorber in the second step effectively reduced the maximum signal amplitude by 30–50 dB in both channels, compared to the frames without the absorber, as shown in **Figure 4**.

- *Step 5: Extraction of mean amplitudes from signal segments.* For each recorded frame, determine the mean values of the maximum amplitudes within the modulation frequency band (around 400 Hz, as shown in **Figure 4**) for the measurement and reference channels, respectively. We assume that these maximum amplitudes can represent the overall amplitude of the THz signal after

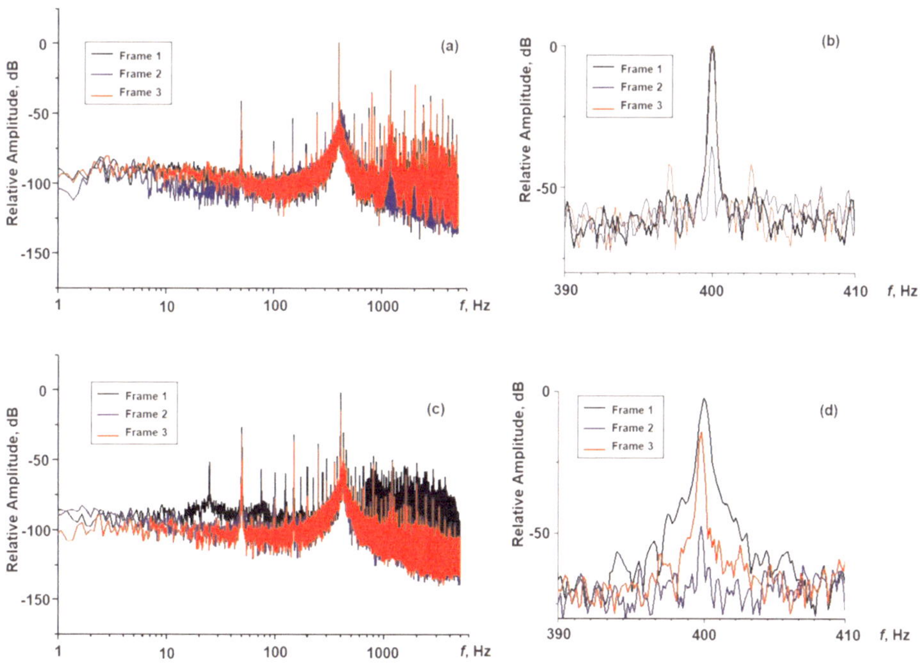

Figure 4.
Spectra of signals from the reference (a, b) and measuring (c, d) channels: (a, c) wide-band analysis; (b, d) narrow-band analysis. Frame 1—Calibration signals without a plate sample in the measuring channel; Frame 2 —Background level without a useful signal; Frame 3—Signals with the measuring plate sample present.

the modulation and demodulation/filtering processes. These values are donated as $S_{m\ \{1,\ 2,\ k\}}$ for the measuring channel and $S_{r\ \{1,\ 2,\ k\}}$ for the reference channel, where $k \geq 3$.

- *Step 6: Calculation of transmission coefficients.* Accounts for laser power fluctuations and corrects for background noise. Calculate the baseline transmission coefficient T_0 and the transmission coefficient with the sample present $T_{(k)}$. To accurately estimate $T_{(k)}$, we propose three different methods:

- *Method 1 (M1): Assessment without background level compensation.*

$$T_{(k)}[\mathrm{dB}] = 10\lg\left(\frac{S_{m(k)}}{S_{m(1)}}\cdot\frac{S_{r(1)}}{S_{r(k)}}\right) = 10\lg\left(\frac{S_{m(k)}}{S_{r(k)}}\cdot\frac{S_{r(1)}}{S_{m(1)}}\right)$$
$$= 10\lg\left(\frac{S_{m(k)}}{S_{r(k)}}\right) - 10\lg\left(\frac{S_{m(1)}}{S_{r(1)}}\right) = 10\lg\left(\frac{S_{m(k)}}{S_{r(k)}}\right) - T_0[\mathrm{dB}]. \quad (32)$$

- *Method 2 (M2): Assessment with background level compensation for only measuring channel.*

$$T_{(k)}[\mathrm{dB}] = 10\lg\left(\frac{S_{m(k)}\frac{S_{r(1)}}{S_{r(k)}} - S_{m(2)}}{S_{m(1)} - S_{m(2)}}\right). \quad (33)$$

- *Method 3 (M3): Assessment with background level compensation for both measuring and reference channels*

$$T_{(k)}[\mathrm{dB}] = 10\lg\left(\frac{S_{m(k)}\frac{S_{r(1)}-S_{r(2)}}{S_{r(k)}-S_{r(2)}} - S_{m(2)}}{S_{m(1)} - S_{m(2)}}\right). \quad (34)$$

- *Step 7: Assessment of material properties.* Use the calculated transmission coefficient T_k to assess the material's transparency. Perform additional measurements with samples of different or same thicknesses to determine key material properties, including the absorption coefficient α and the real (n') and imaginary (n'') parts of the refractive index [9, 31, 32]. Our research specifically targets plane-parallel plates and their absorption properties through transmittance analysis T. To accurately estimate the material's properties, particularly when multiple reflections inside the sample can be neglected (as discussed in Section 2), we employ a method that requires at least two independent measurements. This approach is necessary to determine two unknown parameters: the reflection coefficient from the sample surface (R) and the linear attenuation coefficient (α). The relations can be described using the Fresnel equations when considering normal incidence:

$$\begin{bmatrix} T_1 \\ T_2 \\ \vdots \end{bmatrix} = (1-R)^2 \exp\left(-\boldsymbol{\alpha}\begin{bmatrix} d_1 \\ d_2 \\ \vdots \end{bmatrix}\right). \quad (35)$$

Considering measurement of samples of the same material with two different thicknesses, we derive the following relationships:

$$\begin{aligned}\alpha &= -\frac{\ln T_2 - \ln T_1}{d_2 - d_1} = -\frac{\ln(T_2/T_1)}{d_2 - d_1} \\ &= -\frac{\ln 10^{\frac{T_2[\mathrm{dB}]}{10}} - \ln 10^{\frac{T_1[\mathrm{dB}]}{10}}}{d_2 - d_1} = -\frac{\ln\left(10^{\frac{T_2[\mathrm{dB}]}{10}}/10^{\frac{T_1[\mathrm{dB}]}{10}}\right)}{d_2 - d_1} \\ &= -\frac{\frac{T_2[\mathrm{dB}] - T_1[\mathrm{dB}]}{10}\ln 10}{d_2 - d_1} = -\frac{\ln 10}{10}\frac{T_2[\mathrm{dB}] - T_1[\mathrm{dB}]}{d_2 - d_1}.\end{aligned} \tag{36}$$

Then, taking into account Eqs. (15), (16), (35) and (36), we can determine:

$$n'' = \frac{\alpha}{k} = \frac{\alpha}{2\pi}\lambda. \tag{37}$$

4. Analysis of experimental results

Table 1 shows absorption measurements for various dielectric materials in the submillimeter range. For the same sample with same thickness of plate, differences in experimental series are due to uncontrollable factors such as slight variations in sample orientation. Minor changes in angle can affect signal measurements, especially for anisotropic dielectrics such as leucosapphire, which has a tensor dielectric constant and varies in transparency with rotation. For example, a rotating leucosapphire disk (40 mm diameter, 2 mm thickness) causes fluctuations in the received signal level, impacting its transparency coefficient. Experiments were conducted on dielectrics including mica, foam plastic, fluoroplastic, and leucosapphire [32]. Transparency and attenuation coefficients, with and without background compensation, are detailed in the table. The values labeled M1, M2, and M3 correspond to the transparency coefficients derived from Eqs. (32)-(34). The table also includes specific attenuation coefficients normalized to sample thickness, which are calculated using the following equations:

$$\alpha_{L(k)} = -\frac{T_{(k)}[\mathrm{dB}]}{d_{(k)}}. \tag{38}$$

Materials such as Sitall (glass-ceramic) and Leucosapphire (synthetic sapphire) exhibit notable variation in transparency and attenuation depending on sample thickness and calculation methods. Sitall shows transparency values from −18.08 to −6.67 dB and attenuation coefficients between 4.93 dB/mm and 6.96 dB, while Leucosapphire varies more widely, with transparency ranging from −29.25 to −0.34 dB and attenuation coefficients from 0.12 dB/mm to 14.20 dB. These variations highlight the strong influence of measurement techniques and sample thickness on their optical properties.

Materials such as fused quartz/silica and organic glass exhibit more consistent transparency values. Fused quartz ranges from −7.17 to −6.11 dB, with attenuation coefficients between 1.43 and 2.08 dB/mm. Organic glass shows transparency from −16.46 to −3.88 dB and attenuation coefficients between 2.12 and 6.23 dB/mm. In

Sample type	Composition ratios/ chemical expressions	d (mm)	Transparency coefficient ($T_{(k)}$, – dB)			Specific attenuation coefficient ($\alpha_{L(k)}$, dB/mm)		
			M1	M2	M3	M1	M2	M3
Sitall (glass-ceramic)	SiO_2: ~60–70%; Al_2O_3: ~10–15%; TiO_2, MgO, ZnO: varies based on specific formulation.	2.91	16.98	18.08	18.07	5.63	6.21	6.21
		2.91	16.70	17.71	17.69	4.93	6.09	6.08
		1.94	11.17	11.41	11.41	5.58	5.88	5.88
		1.94	10.91	11.14	11.10	4.77	5.74	5.72
		0.97	6.67	6.75	6.740	6.90	6.96	6.95
		0.97	0.55	0.56	0.38	0.75	0.57	0.40
Leucosapphire (synthetic sapphire)	100% Aluminum oxide (Al_2O_3).	2.26	3.93	3.96	3.97	2.06	1.75	1.76
		2.26	0.62	0.62	0.56	1.49	0.27	0.25
		2.26	0.34	0.34	0.28	1.51	0.15	0.12
		2.06	22.41	29.25	29.24	10.80	14.20	14.20
		2.06	15.30	16.01	15.91	5.43	7.77	7.72
		2.06	11.31	11.56	11.42	2.81	5.61	5.54
		2.06	10.96	11.20	11.13	3.78	5.44	5.40
		2.06	9.57	9.73	9.70	3.70	4.73	4.71
		2.06	9.42	9.58	9.59	4.95	4.65	4.66
		2.06	6.49	6.56	6.55	2.69	3.18	3.18
		2.06	5.64	5.69	5.67	2.12	2.76	2.75
		2.06	3.37	3.39	3.35	0.42	1.65	1.62
Fused quartz/silica	100% Silicon dioxide (SiO_2).	4.3	7.08	7.17	7.17	2.08	1.67	1.67
		4.3	6.11	6.17	6.17	1.54	1.43	1.44
Fiberglass	SiO_2: ~50–60%; CaO, Al_2O_3, B_2O_3, Na_2O, etc.: varies.	17	22.03	27.64	27.59	0.47	1.63	1.62
		4	24.10	31.77	31.82	5.61	7.94	7.96
		1.98	24.27	30.90	30.93	12.02	15.60	15.62
		1.98	22.90	32.34	32.14	10.85	16.33	16.23
Glass	SiO_2: ~70–75%; Na_2O: ~12–15%; CaO: ~10–15%; Minor amounts of Al_2O_3, MgO, etc.	3.47	23.27	37.92	37.29	5.68	10.93	10.75
Organic glass (polymethyl methacrylate, PMMA)	$(C_5O_2H_8)_n$: Repeating units of $C_5O_2H_8$.	2.99	15.67	16.45	16.46	5.80	5.50	5.51
		2.99	15.53	16.28	16.28	5.16	5.45	5.44
		2.42	5.40	5.44	5.17	2.12	2.25	2.14
		2.42	13.45	13.89	13.90	6.06	5.74	5.75
		1.04	3.88	3.90	3.88	3.67	3.75	3.73
		1.04	6.40	6.47	6.48	6.18	6.22	6.23

Sample type	Composition ratios/ chemical expressions	d (mm)	Transparency coefficient ($T_{(k)}$, − dB)			Specific attenuation coefficient ($\alpha_{L(k)}$, dB/mm)		
			M1	M2	M3	M1	M2	M3
Getinax (phenolic paper laminate)	Phenol-formaldehyde resin. Phenolic resin impregnated paper layers.	1.08	13.60	14.06	13.96	12.30	13.02	12.93
		0.18	15.64	16.42	16.42	86.71	91.21	91.22
		0.1	7.3	7.46	7.46	75.76	74.60	74.57
		0.08	3.59	3.62	3.48	103.83	45.25	43.55
Textolite (phenolic laminate)	Phenolic resin with fabric layers	1.17	21.73	26.63	26.62	18.54	22.76	22.75

Table 1.
Transparency coefficients and specific transparency coefficients for different testing sample types.

contrast, Getinax (phenolic paper laminate) displays high attenuation, with coefficients ranging from 12.30 to 103.83 dB/mm, indicating substantial attenuation even in thinner samples, with a minimum value of 43.55 dB/mm for a 0.08 mm thickness.

Overall, **Table 1** reveals that materials with lower specific attenuation coefficients tend to exhibit higher transparency, making them more suitable for THz applications. The observed variations highlight how composition and thickness affect THz radiation interaction, underscoring the need for careful material selection based on transparency and attenuation properties.

Tables 2–4 present estimates of key parameters—linear attenuation coefficient (α) and the imaginary part of the refractive index (n'')—for materials including Sitall, leucosapphire, fiberglass, organic glass, and Getinax. Transparency (T) indicates the amount of THz radiation transmitted through each material, while α shows the extent of signal loss, and n'' provides insight into material interactions with THz waves. These parameters are derived using methods M1, M2, and M3, corresponding to Eqs. (32), (33), and (34), respectively, and are expressed through Eqs. (36) and (37). The tables demonstrate how different calculation methods impact transparency measurements and offer valuable insights into material performance for THz applications. Detailed analyses of these tables are as follows.

Table 2 presents estimates of material properties using the M1 method (Eq. (32)) for THz frequencies from 790 to 1000 GHz. Materials such as Sitall, fiberglass, and organic glass show high negative T (dB) values, indicating excellent THz transparency. Conversely, Getinax, a phenolic paper laminate, shows much lower T values at 0.1 and 0.08 mm thicknesses, reflecting high attenuation. Getinax has a notably high α value of 75.76 mm^{-1}, indicating severe THz attenuation and absorption, unlike fiberglass with a low α of −1.63 mm^{-1}, suggesting minimal attenuation. Sitall and organic glass also show lower α values, confirming their superior transparency compared to Getinax. Additionally, Getinax's high n'' values (up to 0.0746, or 7.46%) indicate substantial absorption and reflection of THz radiation, contributing to its high attenuation. In contrast, materials such as Leucosapphire and organic glass have lower n'' values (1.75–5.80%), suggesting reduced absorption and better THz transparency. These results underscore the importance of choosing materials with lower n'' and α values for effective THz transmission and minimal signal loss.

Sample type	d (mm)	$T_{(k)}$ (dB)	$T_{(k)}$ (%)	$\alpha_{(k)}$ (mm^{-1})	n'' (%)
Sitall	2.91	−16.98	2.0045	1.3792	0.6585–0.8336
	1.94	−11.17	7.6384		
Leucosapphire	2.26	−3.93	40.4576	0.6447	0.3078–0.3897
	2.06	−3.37	46.0257		
Fiberglass	4	−24.10	0.3890	−0.0194	(−0.0117) – (−0.0093)
	1.98	−24.27	0.3741		
Organic glass	2.99	−15.67	2.7102	0.8968	0.4282–0.5420
	2.42	−13.45	4.5186		
Getinax	0.1	−7.3	18.6209	43.9794	20.9986–26.5805
	0.08	−3.48	44.8745		

Table 2.
Estimation of material properties using the M1 method for transparency coefficient (T) calculation at 790–1000 GHz laser frequencies.

Sample type	d (mm)	$T_{(k)}$ (dB)	$T_{(k)}$ (%)	$\alpha_{(k)}$ (mm^{-1})	n'' (%)
Sitall	2.91	−18.08	1.5560	1.5833	0.7560–0.9569
	1.94	−11.41	7.2277		
Leucosapphire	2.26	−3.96	40.1791	0.6562	0.3133–0.3966
	2.06	−3.39	45.8142		
Fiberglass	4	−31.77	0.0665	0.0992	0.0474–0.0599
	1.98	−30.90	0.0813		
Organic glass	2.99	−16.45	2.2646	1.0341	0.4938–0.6250
	2.42	−13.89	4.0832		
Getinax	0.1	−7.46	17.9473	44.2096	21.1085–26.7197
	0.08	−3.62	43.4510		

Table 3.
Estimation of material properties using the M2 method for transparency coefficient (T) calculation at 790–1000 GHz laser frequencies.

It is worth mentioning that the negative value of n'' for fiberglass is unusual and warrants clarification. Typically, n'' represents the material's absorption coefficient and is expected to be non-negative. A negative n'' suggests that there might be an issue with the measurement or calculation process, as it implies that the material is generating energy rather than absorbing it, which is not physically plausible. This negative value could arise from measurement errors, inaccuracies in the background noise correction, or incorrect assumptions in the modeling. Therefore, correct that background noise can significantly affect the accuracy of transparency measurements. Methods M2 and M3 are designed to address this issue.

Sample type	d (mm)	$T_{(k)}$		$\alpha_{(k)}$ (mm^{-1})	n'' (%)
		(dB)	(%)		
Sitall	2.91	−18.07	1.5596	1.5810	0.7548–0.9555
	1.94	−11.41	7.2277		
Leucosapphire	2.26	−3.97	40.0867	0.7138	0.3408–0.4314
	2.06	−3.35	46.2381		
Fiberglass	4	−31.82	0.0658	0.1015	0.0484–0.0613
	1.98	−30.93	0.0807		
Organic glass	2.99	−16.46	2.2594	1.0341	0.4938–0.6250
	2.42	−13.90	4.0738		
Getinax	0.1	−7.46	17.9473	44.5550	21.2735–26.9284
	0.08	−3.59	43.7522		

Table 4.
Estimation of material properties using the M3 method for transparency coefficient (T) calculation at 790–1000 GHz laser frequencies.

Table 3 shows T for various materials using Method M2 (refer to Eq. (33)), which compensates for background noise in the measuring channel. Sitall and organic glass have high transparency with coefficients of −18.08 and − 16.45 dB, respectively. In contrast, Getinax exhibits significantly lower transparency with coefficients of −7.46 and − 3.62 dB, indicating high attenuation. α reflects how much THz radiation is absorbed. Getinax has the highest α at 44.2096 mm^{-1}, showing very high attenuation. Fiberglass, with a lower α of 0.0992 mm^{-1}, demonstrates high transparency despite its transparency coefficient. Different from **Table 2**, here fiberglass shows a positive low n'' of 0.0474–0.0599%, indicating minimal absorption. Getinax has a much higher n'' (21.1085–26.7197%), aligning with its high attenuation. Reflectance values vary among materials, with fiberglass and organic glass exhibiting higher reflectance, which reduces their overall transparency. However, due to the high α for Sitall and Getinax, Eq. (35) is not suitable for deriving their reflectance values.

Table 4 presents T for various materials using Method M3 (refer to Eq. (34)), which includes background noise compensation for both measuring and reference channels. Sitall and organic glass show high transparency with coefficients of −18.07 and − 16.46 dB, respectively. In contrast, Getinax has significantly lower transparency coefficients of −7.46 and − 3.59 dB, indicating high attenuation. Getinax exhibits the highest α value at 44.5550 mm^{-1}, indicating very high attenuation. Fiberglass, with a lower α of 0.1015 mm^{-1}, demonstrates higher transparency despite its transparency coefficient. Fiberglass has a low n'' (0.0484–0.0613%), indicating minimal absorption. In contrast, Getinax has a significantly higher n'' (21.2735–26.9284%), consistent with its high attenuation. Reflectance values further highlight material differences, with fiberglass and organic glass exhibiting higher reflectance, which impacts their effective transparency.

The comparison of **Tables 2-4** indicates that while transparency and attenuation values are similar across methods, method M3 provides the most accurate and reliable results. This is due to its comprehensive background noise compensation, in the both measuring and reference channels, leading to more precise measurements. These findings align with our previous research [32], which highlighted the importance of background noise compensation in enhancing measurement accuracy.

5. Critical discussion

This study provides a detailed analysis of the transparency and attenuation properties of various dielectric materials at THz frequencies using a 2-channel measuring system with a THz gas laser. The 400 Hz modulation and demodulation techniques are creatively applied in the measurement process. However, several limitations affect the accuracy and reliability of the results:

1. *Sample's thickness sensitivity:* Variations in results for materials such as Leucosapphire and Getinax highlight the need for precise control and measurement of sample thickness.

2. *Sample's composition error:* Variations in material composition, such as the ratios in Sitall or crystal orientation in Leucosapphire, introduce significant variability in THz interaction, complicating result interpretation.

3. *Methodological discrepancies:* Differences among methods M1, M2, and M3 reveal the importance of background noise compensation, indicating the need for standardized methods to improve data consistency.

4. *High attenuation challenges:* The high attenuation observed in materials like Getinax suggests system limitations in accurately capturing properties at extreme attenuation levels.

5. *Measurement errors:* Negative values in the n'' for materials like fiberglass point to potential errors in measurement or data processing.

6. *Calibration and validation:* Improved calibration techniques and method validation under varied conditions are necessary to enhance measurement accuracy.

7. *Environmental factors:* The study does not consider environmental conditions such as temperature or humidity, which could affect measurements and THz interaction with materials.

Despite these limitations, method M3 offers the most accurate results due to comprehensive noise compensation. Future work should validate these methods further and integrate recent advancements in THz research to enhance measurement accuracy and applicability. Moreover, recent advancements in THz research, including graphene-based THz devices [37], metamaterials [38], and time-domain THz spectroscopy (THz-TDS) [39], underscore the necessity for continued development in measurement techniques and material characterization. Integrating these innovations could significantly improve measurement accuracy and expand the practical applications of THz technologies.

6. Conclusion

This study evaluates dielectric materials for terahertz (THz) applications using a specialized THz measuring system with a 2-channel setup, 400 Hz amplitude modulator, and gas laser. Analysis of the 400 ± 10 Hz narrowband filtered (demodulated) signal levels, which represents the overall THz signal, revealing that materials such as Sitall and Leucosapphire, with lower specific attenuation coefficients, provide greater transparency and are thus more suitable for THz technologies. In contrast, materials such as Getinax, which exhibit high attenuation even in thinner samples, are less effective in THz environments.

The research underscores the importance of background noise compensation for accurate and reliable measurements. Techniques such as signal modulation with shutters and continuous signal level monitoring significantly improve measurement precision, aligning with previous studies that emphasize minimizing measurement errors.

The data obtained are crucial for estimating the complex refractive indices of solid dielectrics within the submillimeter wavelength range, supporting the advancement of THz technologies. Future research should focus on optimizing material selection and refining measurement systems to enhance the precision and reliability of THz applications. Employing advanced signal modulation techniques and extending the dynamic range of attenuation measurements will be vital for achieving accurate characterization and producing reliable data in upcoming studies.

Acknowledgements

The work by Olena Kryvenko and Vladyslav Lutsenko was supported by the National Academy of Sciences of Ukraine.

Conflict of interest

The authors declare no conflict of interest.

Thanks

The authors would like to thank IntechOpen for their support in publishing this article, particularly for the assistance of Ms. Marica Novakovic from the editorial office. We also extend our gratitude to Prof. Vladimir Gavrikov and Dr. Oleh Polikarpov for their valuable advice, insightful discussions, and technical support with the laser generation equipment. Additionally, we are grateful to the academic editor, Professor Jinfeng Li, for his valuable suggestions, which significantly contributed to the enhancement of the manuscript.

Author details

Yiyang Luo[1,2*], Olena Kryvenko[1], Vladyslav Lutsenko[1] and Vitalii Yakovliev[3,4]

1 O. Ya. Usikov Institute for Radiophysics and Electronics of the NAS of Ukraine, Kharkiv, Ukraine

2 V.N. Karazin Kharkiv National University, Kharkiv, Ukraine

3 Institute of Radio Astronomy of the National Academy of Sciences of Ukraine, Kharkiv, Ukraine

4 National Aerospace University M.E. Zhukovsky (KhAI), Kharkiv, Ukraine

*Address all correspondence to: yiyangluo@163.com

IntechOpen

© 2025 The Author(s). Licensee IntechOpen. This chapter is distributed under the terms of the Creative Commons Attribution License (http://creativecommons.org/licenses/by/4.0), which permits unrestricted use, distribution, and reproduction in any medium, provided the original work is properly cited.

References

[1] Smye SW, Chamberlain JM, Fitzgerald AJ, Berry E. The interaction between terahertz radiation and biological tissue. Physics in Medicine and Biology. 2001;**46**(3):R101. DOI: 10.1088/0031-9155/46/9/201

[2] Dudovich N, Oron D, Silberberg Y. Single-pulse coherently controlled nonlinear Raman spectroscopy and microscopy. Nature. 2002;**418**(6894):512-514. DOI: 10.1038/nature00933

[3] Cole BE, Williams JB, King BT, Sherwin MS, Stanley CR. Coherent manipulation of semiconductor quantum bits with terahertz radiation. Nature. 2001;**410**(6825):60-63. DOI: 10.1038/35065032

[4] Fitzgerald AJ, Berry E, Zinovev NN, Walker GC, Smith MA, Chamberlain JM. An introduction to medical imaging with coherent terahertz radiation. Physics in Medicine and Biology. 2002;**47**(21):67-84. DOI: 10.1088/0031-9155/47/7/201

[5] Kent G. Nondestructive permittivity measurement of substrates. IEEE Transactions on Instrumentation and Measurement. 1996;**45**(1):102-106. DOI: 10.1109/19.481319

[6] Janezic MD. Full-wave analysis of a split-cylinder resonator for nondestructive permittivity measurement. IEEE Transactions on Microwave Theory and Techniques. 1999;**47**(12):2014-2020. DOI: 10.1109/22.795077

[7] Hammler J, Gallant A. Balocco. Simple de-embedding and simulation technique to find permittivity with a THz vector network analyser. In: Proceedings of the IRMMW-THz. Piscataway, NJ: IEEE; 2015. pp.1-2. DOI: 10.1109/IRMMW-THz.2015.7327822

[8] James J. A Student's Guide to Fourier Transforms. New York (NY): Cambridge University Press; 2011. ISBN: 9780511762307

[9] Zandonella C. T-ray specs. Nature. 2003;**424**(6945):721-722. DOI: 10.1038/424721a

[10] Pavlovskii VV, Gundareva II, Volkov OY, et al. Extension of the frequency range of Josephson impedance spectroscopy. Journal of Communications Technology and Electronics. 2013;**58**(9):951-955. DOI: 10.1134/S1064226913090106

[11] Snezhko A, Gubankov V. Coupling of bicrystal Josephson junction and planar resonator in the sub-THz frequency range. Nonlinear World. 2015;**13**(2):46-48

[12] Miles RE, Harrison P, Lippens D, editors. Terahertz sources and systems. In: NATO Science Series, II, 27. Kluwer Academic Publishers; 2001. DOI: 10.1007/978-94-010-0824-2

[13] Mittleman DM, editor. Sensing with terahertz radiation. In: Springer Series in Optical Sciences, 85. Springer-Verlag; 2003. DOI: 10.1007/978-3-540-45601-8

[14] Van der Weide D. Applications and outlook for electronic terahertz technology. Optics & Photonics News. 2003;**14**(4):48-53. DOI: 10.1364/OPN.14.4.000048

[15] Svelto O. Principles of lasers. In: Springer Book Archive. 1st ed. New York, NY: Springer; 1989. p. XIII-494. DOI: 10.1007/978-1-4615-7670-9

[16] Hu VV, Zhang X-C, Auston DH. Terahertz radiation induced by subband-

gap femtosecond optical excitation of GaAs. Physical Review Letters. 1991; **67**(22):2709-2712. DOI: 10.1103/PhysRevLett.67.2709

[17] Kersting R, Unterrainer K, Strasser G, Kauffmann HF, Gornik E. Few-cycle THz emission from cold plasma oscillations. Physical Review Letters. 1997;**79**(15):3038-3041. DOI: 10.1103/PhysRevLett.79.3038

[18] Matsuura S, Tani M, Sakai K. Generation of coherent terahertz radiation by photomixing in dipole photoconductive antennas. Applied Physics Letters. 1997;**70**(5):559-561. DOI: 10.1063/1.118337

[19] Kroug M, Cherednichenko S, Merkel H, Kollberg E, Voronov B, Gol'tsman G, et al. NbN hot electron bolometric mixers for terahertz receivers. IEEE Transactions on Applied Superconductivity. 2001;**11**(1):962-965. DOI: 10.1109/77.919508

[20] Burke PJ, Schoelkopf RJ, Prober DE, Skalare A, Karasik S, Gaidis MC, et al. Mixing and noise in diffusion and phonon cooled superconducting hot-electron bolometers. Journal of Applied Physics. 1999;**85**(3):1644-1653. DOI: 10.1063/1.369299

[21] Karasik BS, McGrath WR, Gershenson ME, Sergeev AV. Photon-noise-limited direct detector based on disorder-controlled electron heating. Journal of Applied Physics. 2000;**87**(11): 7586-7588. DOI: 10.1063/1.373026

[22] Crow TW, Mattauch RJ, Weikle RM, Bhapkar UV. Terahertz GaAs devices and circuits for heterodyne receivers. In: Shur M, editor. Compound Semiconductor Electronics. Singapore: World Scientific Publishing; 1996. pp. 209-246. DOI: 10.1142/9789812830821_0005

[23] Marazita SM, Bishop WL, Hesler JL, Hui K, Bowen WE, Crowe TW. Integrated GaAs Schottky mixers by spin-on-dielectric wafer bonding. IEEE Transactions on Electron Devices. 2000; **47**(5):1152-1157. DOI: 10.1109/16.842956

[24] Reichertz LA, Strojnik M, Beeman JW, Cardozo BL, Haegel NM, Haller EE, et al. GaAs BIB photodetector development for far-infrared astronomy. In: SPIE Proceedings. Denver (Colorado). Bellingham, WA: SPIE; pp. 231-238. DOI:10.1117/12.560291

[25] Sahin S, Nahar NK, Sertel K. Dielectric properties of low-loss polymers for mmW and THz applications. Journal of Infrared, Millimeter, and Terahertz Waves. 2019; **40**:557-573. DOI: 10.1007/s10762-019-00584-2

[26] Nazarov MM, Cherkasova OP, Shkurinov AP. Study of dielectric function of glucose and albumin aqueous solutions by pulsed terahertz spectroscopy. Quantum Electronics. 2016;**46**(6):488-495. DOI: 10.1070/QEL16107

[27] Mei HW, Li L, Yin F, Hu W, Liu J, Wang L. Study on detecting dielectric properties of typical electrical insulation materials by terahertz wave spectroscopy. IEEE Access. 2019;**7**: 129190-129200. DOI: 10.1109/ACCESS.2019.2941798

[28] Liu X, Fan K, Shadrivov IV, Padilla WJ. Experimental realization of a terahertz all-dielectric metasurface absorber. Optics Express. 2017;**25**(1): 191-200. DOI: 10.1364/OE.25.000191

[29] Park SJ, Cunningham J. Determination of permittivity of dielectric analytes in the terahertz frequency range using split ring

resonator elements integrated with on-chip waveguide. Sensors. 2020;**20**(15): 4264. DOI: 10.3390/s20154264

[30] Sarkar S, Saha D, Banerjee S, Mukherjee A, Mandal P. Broadband terahertz dielectric spectroscopy of alcohols. Chemical Physics Letters. 2017; **684**:278-283. DOI: 10.1016/j.cplett.2017.04.026

[31] Ionin AA, Ivanov S, Kotkov AA, Kozlov AY, Seleznev LV. Multicomponent gas analysis with first-overtone CO laser. In: Proc. SPIE 4760, High-Power Laser Ablation IV. Bellingham, WA: SPIE; 2002. pp.1078-1087. DOI: 10.1117/12.482071

[32] Gavrikov VK, Krivenko EV, Lutsenko VI, Popov IV, Luo Y, Polikarpov OV, et al. Measuring complex for study the characteristics of dielectrics in the submillimeter wavelength range. In: 2020 IEEE Ukrainian Microwave Week (UkrMW), Kharkiv, Ukraine. Piscataway, NJ: IEEE; 2020. pp. 841-844. DOI: 10.1109/UkrMW49653.2020.9252797

[33] Yunovich AE, Lukyanov FA. Optical Phenomena in Semiconductors: Special Practice Task. Moscow: Moscow State University; 2007. 49 p. [In Russian]

[34] Jitsuno T, Uno K. CO_2 lasers. In: Sugioka K, editor. Handbook of Laser Micro- and Nano-Engineering. Cham: Springer; 2021. DOI: 10.1007/978-3-319-69537-2_56-1

[35] Grover HS, Dawson F. Minimizing late currents in pulsed power gas laser systems. IEEE Transactions on Plasma Science. 2021;**49**(9):2946-2953. DOI: 10.1109/TPS.2021.3098759

[36] Shumakova V, Heckl OH. A short guide to recent developments in laser-based gas phase spectroscopy, applications, and tools. APL Photonics. 2024;**9**:010803. DOI: 10.1063/5.0167683

[37] Otsuji T, Boubanga-Tombet SA, Fukidome M, Sano E. Graphene-based terahertz devices: Advances and future prospects. IEEE Transactions on Terahertz Science and Technology. 2021; **11**(5):405-420. DOI: 10.1109/TTHZ.2021.3064532

[38] Withayachumnankul W, Abbott D. Metamaterials in the terahertz regime. IEEE Photonics Journal. 2009;**1**(2): 99-118. DOI: 10.1109/JPHOT.2009.2034969

[39] Jepsen PU, Cooke DG, Koch M. Terahertz spectroscopy and imaging – Modern techniques and applications. Laser & Photonics Reviews. 2011;**5**(1): 124-166. DOI: 10.1002/lpor.201000011

Chapter 5

Unveiling Precision: Harnessing Terahertz Technology for Precise Margin Detection

Jyotirmayee Dash, Sidharth Raj, Arun Jana, Vijay Pillai, Akshita Singh and Nisheena Raghavan

Abstract

In the pursuit of improved accuracy in medical diagnostics and treatment, Terahertz (THz) technology has emerged as a powerful tool for precise margin detection, particularly in oncology. This chapter explores the cutting-edge advancements in THz imaging and spectroscopy, emphasizing their potential to delineate tumor margins with unprecedented precision. THz waves, with their unique interaction with biological tissues, offer a non-invasive and highly sensitive approach to distinguishing between healthy and malignant tissues. The ability of THz technology to provide real-time, label-free imaging makes it a promising candidate for intraoperative use, potentially reducing the rates of incomplete resection and recurrence. We discuss the principles underlying THz imaging, its integration into clinical workflows, and the challenges that must be addressed to fully realize its potential. The purpose of this chapter is to highlight the transformative impact of THz technology on margin detection, paving the way for more accurate, personalized, and effective surgical interventions.

Keywords: Terahertz imaging, Terahertz spectroscopy, breast cancer, oral cancer, Artificial Intelligence, CNN, ANN

1. Introduction

The landscape of modern medicine is increasingly defined by its pursuit of precision, where every decision can have profound implications for patient outcomes. Among the various challenges faced by clinicians, the accurate identification and delineation of tumor margins during surgical procedures stand as one of the most critical [1]. For patients diagnosed with cancer, particularly in cases of breast cancer and oral cancer, the removal of malignant tissue through surgery remains a cornerstone of treatment. However, the ability to ensure complete resection without damaging surrounding healthy tissue is paramount. In breast cancer, clear margins are essential to reduce the risk of local recurrence and improve long-term survival rates. Similarly, in oral cancer, precise margin detection is crucial to prevent functional impairments and improve cosmetic outcomes. Incomplete removal of cancerous tissue can lead to local recurrence, metastasis, and ultimately, diminished survival rates [2]. Conversely, the excision of excessive healthy tissue can result in unnecessary loss of function and reduced quality of life [3]. It is within this delicate balance that the need for precise margin detection becomes evident.

Traditional imaging modalities, such as magnetic resonance imaging (MRI), computed tomography (CT), and ultrasound, have long been employed to assist surgeons in navigating this complex terrain [4]. These technologies, while invaluable in providing preoperative and intraoperative guidance, often face limitations in spatial resolution, sensitivity, and the ability to differentiate between subtle variations in tissue types. For instance, MRI, with its superior soft tissue contrast, may not always clearly distinguish between tumor and surrounding inflamed or fibrotic tissues [5]. Similarly, CT, while excellent for structural imaging, lacks the necessary contrast at the molecular level to differentiate between malignant and benign tissues with absolute certainty. As a result, there is a growing recognition within the medical community of the need for innovative imaging techniques that can provide enhanced specificity and sensitivity at the surgical site [6].

In this context, THz technology has emerged as a promising candidate, offering a novel approach to margin detection that addresses many of the shortcomings of conventional methods [7]. Terahertz radiation, which occupies the spectral range between microwaves and infrared light, has unique properties that make it particularly well-suited for biomedical imaging [8]. THz waves are non-ionizing, meaning they do not carry the risk of damaging tissues or DNA, as is the case with X-rays. Moreover, they are highly sensitive to the water content, tissue density, and molecular composition of biological tissues, allowing for the detection of subtle differences that might not be visible with other imaging modalities [9]. One of the most compelling features of THz imaging is its ability to provide label-free, real-time imaging of tissues [10]. This capability is especially important in a surgical setting, where time is of the essence, and decisions must be made quickly to minimize the patient's time under anesthesia and reduce the risk of complications. The interaction of THz waves with biological tissues results in unique spectral signatures that can be used to distinguish between healthy and malignant tissues with high precision. For example, cancerous tissues often have higher water content and different molecular compositions than surrounding healthy tissues, which can be detected by THz imaging systems. This makes THz technology particularly valuable for identifying tumor margins during surgery, where the distinction between cancerous and non-cancerous tissue can be challenging even for experienced surgeons. Magda El-Shenawee et al. have demonstrated the potential of differentiating tissue types of excised breast cancer tumors [11]. They have utilized the frequency range of 0.1 to 3.5 THz for detecting tissue types including ductal carcinomas, fibro-grandular, and fat tissues. A small piece of freshly excised tissue has been measured via transmission spectroscopy as shown in **Figure 1**.

The results show significant difference between cancer and fat and less clear differentiation between cancer and fibro-grandular tissue. Tahereh et al. [12] have brought out many novel properties of Terahertz radiation for medical diagnostics and treatments with considerable advantages. The authors have shown the special features of the Terahertz radiation in **Figure 2**.

Yang et al. [13] have described the recent developments of THz imaging technique for breast cancer detection. The authors emphasize THz imaging to distinguish cancerous and peritumoral tissue from normal tissue with clear boundaries. Numerous

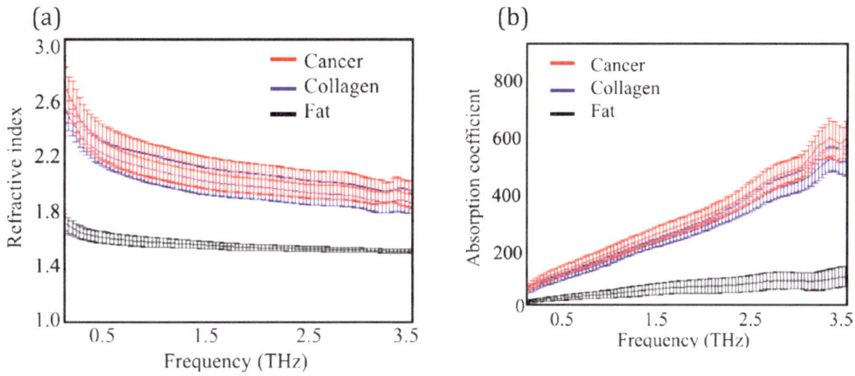

Figure 1.
Mean spectroscopy data for all fresh human cancer and healthy breast tissues: (a) The transmission refractive indices and (b) transmission absorption coefficient of cancer, collagen, and fat [11].

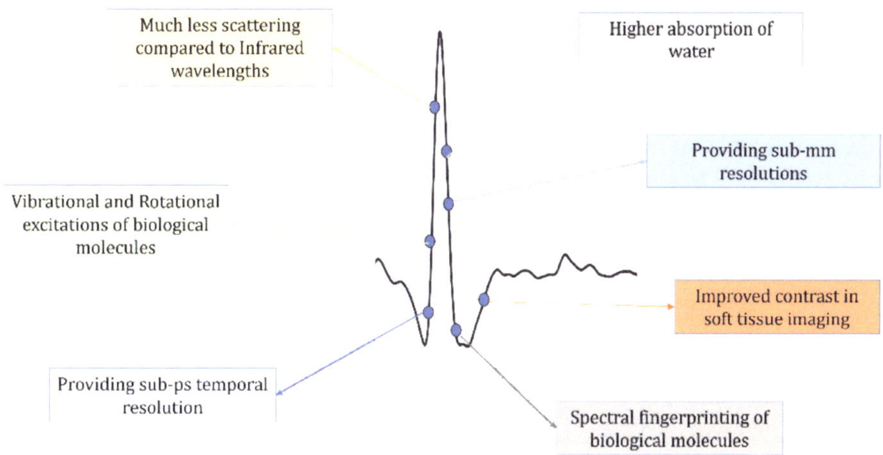

Figure 2.
Special features of THz radiation.

chemometric methods such as Principal Component Analysis (PCA), Linear Discriminant Analysis (LDA), Decision Tree (DT), Genetic Algorithm (GA) have been utilized to improve THz imaging [13].

However, until recently, clinical trials of THz devices on fresh tissue samples on human tissues are still limited due to the unavailability of these tissues outside the hospital and inaccessibility to real hospital environments. A significant milestone in the clinical application of THz technology has been achieved by TeraLumen Solutions, which has developed a compact continuous wave (CW) THz system [14]. This system has been successfully deployed in a clinical environment, where it has been used to clinically validate more than 170 fresh tissue samples [15]. The clinical validation process was conducted with prior institutional ethical clearance from the Institutional Review Board (IRB), ensuring that all ethical guidelines were adhered to. Furthermore, informed consent was obtained from each patient involved in the study, underscoring the ethical rigor of this pioneering work. The results of these clinical trials, included in this chapter, provide compelling evidence of the potential of THz technology in real-world clinical settings.

The integration of THz technology into the surgical workflow represents a significant advancement in the field of intraoperative imaging. As THz systems become more refined and compact, their adoption in operating rooms could transform the way surgeons approach margin detection. The potential benefits extend beyond oncology to other fields where precise tissue characterization is crucial, such as dermatology, ophthalmology, and neurology. However, the journey from concept to clinical application is not without challenges. The development and optimization of THz imaging systems, as discussed in this chapter, emphasize the need for further advancements in data processing algorithms to enhance image quality and support clinical decision-making. The interpretation of THz data necessitates sophisticated algorithms and software tools that can provide actionable insights in real time.

In this chapter, we explored the fundamental principles of THz imaging, from the underlying physics to the engineering of THz systems. We will review the current state of research in THz-based margin detection, highlighting key studies that demonstrate its potential in clinical settings. Moreover, we will discuss the technical and practical challenges that must be overcome to fully realize the benefits of THz technology in margin detection. These challenges include optimizing the spatial resolution of THz imaging, shape deformation between THz image and histopathology image, scattering near the edge, and developing standardized protocols for its use in different types of surgeries. Furthermore, we will consider the economic and regulatory factors that influence the adoption of new technologies in healthcare, with a focus on the path to clinical translation for THz imaging.

2. Principle of THz imaging

Over the past two decades, the development and deployment of THz imaging systems have seen remarkable progress, driven by significant improvements in the production and detection of THz radiation. These advancements have led to the widespread availability of commercial THz sources, which are broadly categorized into CW and pulsed sources. CW THz sources can be further divided into photonic sources, photomixers, and electronic sources. Pulsed THz sources are mainly photoconductive antennas and optical rectification, as shown in **Figure 3**. Similarly, the detectors are also categorized into coherent and incoherent detectors as shown in **Figure 4**.

A pulsed THz systems can produce broadband emissions, capturing both intensity and phase information, thus offering more detailed insights into the target. The schematic of THz pulsed system is shown in **Figure 5**. Techniques used in pulsed systems include ultrafast switching of photoconductive antennas, rectification of optical pulses in crystals, and carrier tunneling in coupled double quantum well structures. Among these, the use of photoconductive antennas with femtosecond lasers is the most established method, despite being expensive. Nonetheless, the advances in THz technology have led to the possible use of THz radiation-based technology as an insightful medical tool [16]. They have revealed the THz Pulsed Imaging (TPI)-based THz pulsed radiation detection methods and their potential applications in the healthcare domain. The numerous advantages of THz pulsed radiation-based imaging were outlined. Utilizing the THz radiation properties, typical deployment studies of TPI for in vivo and ex vivo cancer results were laid out. In order to improve penetration depth and sensing capabilities for biological applications, they have also addressed the limits of THz imaging technology. The authors observed the swift advancements in THz imaging technology and its growing promise as a medical

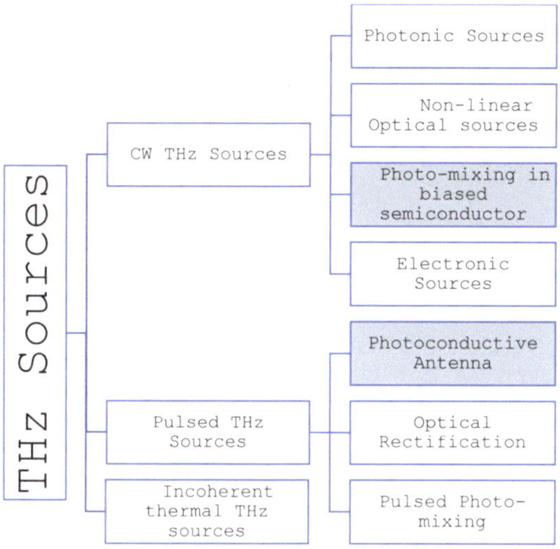

Figure 3.
Various types of THz sources used to build THz systems.

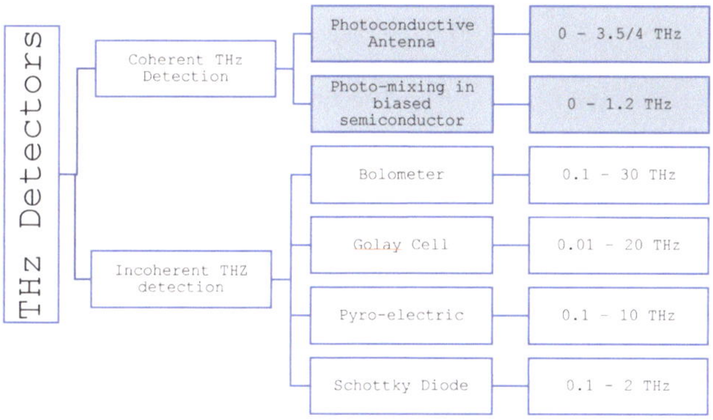

Figure 4.
Various types of coherent and incoherent detectors used to detect THz radiation and their achievable bandwidth.

imaging modality. Nevertheless, they indicated that further focus on the application of PEAs and technological advancements is necessary for the clinical implementation of the TPI system.

In this chapter, we are focusing on CW THz systems, as it has been used for the clinical trial. CW THz system is known for its ability to emit single or multiple discrete frequencies and generate THz radiation using techniques such as photomixing, free-electron lasers, and quantum cascade lasers. These systems are relatively straightforward in design, allowing for compact and cost-effective solutions. However, their narrow spectral range can limit the amount of information they can provide. A comprehensive examination of CW THz point-by-point scanning techniques was presented, shedding light on the state and applications of CW THz imaging methods for biomedical materials [17]. In particular, the investigation focused on the biological applications of CW THz imaging, offering a summary of the current research, advancements, benefits, and drawbacks that hinder the widespread adoption of this technology in medicine. The prospects for future development in CW

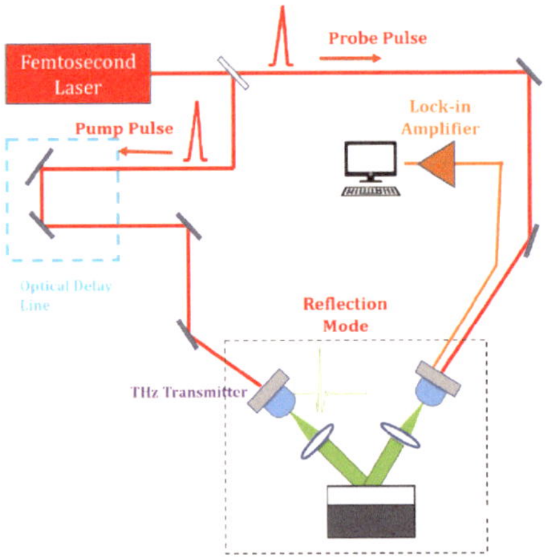

Figure 5.
Schematic of THz pulsed system.

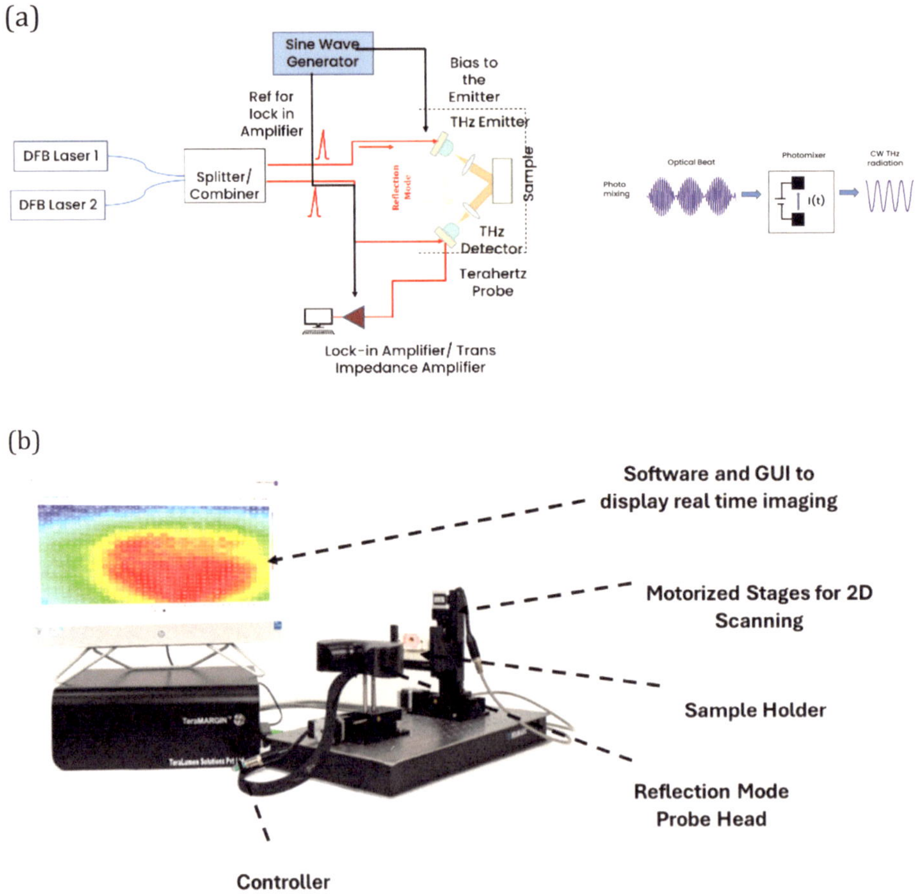

Figure 6.
(a) A schematic of a CW system (b) Device developed by TeraLumen for the clinical trial of fresh tissue samples.

THz imaging were also discussed, particularly in the context of its potential use for biomedical samples [17].

The CW THz imaging system, developed by TeraLumen Solutions, spans from 0.1 to 0.8 THz with a dynamic range of 40 to 50 dB. The system uses two Distributed Feedback Lasers with a center wavelength at 780 nm and LT-GaAs-based Photomixers for the generation and detection of THz signal. Complete electronics including the bias and amplifier circuit has been developed by TeraLumen. The schematic of the system is shown in **Figure 6(a)**, which uses a photomixing technique as shown in the figure. The packaged system that has been utilized for the measurement of tissue samples is shown in **Figure 6(b)**.

As shown in **Figure 6(b)**, the device has two important components, controller and probe head. The controller holds all the electronics such as the bias circuit, power distribution board, and amplifier circuit. The reflection probe head houses the THz emitter and detector along with the related optics to focus the beam. The sample holder is customized for tissue imaging. The tissue imaging is carried out by automated movement of the motorized stages.

3. Revolutionizing early detection: THz imaging at the frontline of cancer

THz imaging and spectroscopy have garnered attention for their potential in cancer diagnostics, particularly for breast and oral cancers. THz waves are non-ionizing, providing high spatial resolution and sensitivity to the water content and molecular composition of tissues. For instance, malignant tissues generally have higher water content and distinct THz signatures compared to healthy tissues. These properties make THz techniques well-suited for distinguishing between cancerous and healthy tissues, which is crucial for accurate margin detection during surgery.

3.1 THz imaging and spectroscopy for breast cancer margin detection

THz imaging has been extensively studied for ex vivo breast tissue samples, where it has demonstrated the ability to differentiate between normal, benign, and malignant tissues [18]. While in vivo applications are more challenging due to factors like the limited penetration depth of THz waves and tissue scattering, there has been progress in using THz imaging during breast-conserving surgeries. The goal is to assess tumor margins in real time to ensure the complete excision of cancerous tissues while preserving as much healthy tissue as possible. Some studies have reported promising results, suggesting that THz imaging could reduce the rates of re-excision.

For instance, recent studies investigated the use of THz imaging combined with deep learning techniques to improve the classification of breast cancer tissues [19]. The authors focus on analyzing freshly excised breast cancer tissue samples from mice, applying a method known as wavelet synchro-squeezed transformation (WSST) to convert THz data into spectrograms. These spectrograms are then processed through deep convolutional neural networks (CNNs) for more accurate tissue classification. The study demonstrates that the deep learning approach significantly enhances the ability to distinguish between cancerous and non-cancerous tissues, outperforming traditional statistical methods. This improvement is particularly relevant for segmenting areas where cancerous tissue intersects with other tissue types, such as muscle. The authors highlight the potential for this technology to provide near real-time analysis during surgical procedures, which could lead to quicker and more accurate decision-making in breast cancer surgeries. On the other hand, the utilization of THz imaging technology for breast cancer diagnosis has been documented, and this includes applications in addition to the instances in **Table 1**.

Nonetheless, most of the publications have utilized the THz-TDS system for the analysis. We at TeraLumen Solutions have developed a low-cost, compact affordable, and easily accessible CW THz system, which is shown in **Figure 6** and validated with 86 breast tissue samples and 110 Head and Neck cancer tissue samples.

3.2 Fresh tissue sample measurement protocol

As mentioned in the above paragraph, the authors of this chapter have carried out THz measurements of more than 170 fresh tissue samples using the CW system developed by TeraLumen Solutions. The procedure of the fresh sample measurement is as follows. Using sterile forceps, the fresh tumor sample was transferred from the Dulbecco's Modified Eagle's medium (DMEM) to a sterile petri dish. A total of 75% ethanol was sprayed on the petri dish, forceps, and gloves to maintain aseptic

Study	THz system	Key findings
Wallace et al. [20]	TPI Scanner (TeraView Ltd., Cambridge, UK)	Successfully identified the extent of BCC in both in vivo and excised tissues; delineated tumor margin.
Fitzgerald et al. [21]	TPI Scanner (TeraView Ltd., Cambridge, UK)	Capable of detecting invasive breast carcinoma and ductal carcinoma.
Ashworth et al. [18]	Portable THz Pulsed Transmission Spectrometer	THz Pulsed Spectroscopy and TPI could differentiate between healthy adipose, fibrous, and cancerous breast tissues.
Chen et al. [22]	CW THz Near-Field Microscopy Transmission Imaging	Able to distinguish breast tumors from normal tissue without H&E staining, achieving a resolution of 240 μm.
Joseph et al. [23]	CW THz Transmission Imaging	Demonstrated clear contrast between cancerous and normal tissues with a spatial resolution of 390 μm at 1.4 THz and 490 μm at 1.6 THz.
Peter et al. [24]	CW THz Imaging Mode	Observed absolute refractive index differences between tissue samples.
Bowman et al. [25]	TPS Spectra 3000 Model	Capable of detecting heterogeneous samples with a thickness as small as 10 μm.
Bowman et al. [26]	TPS Spectra 3000 System	Provided higher resolution and more apparent margin differences between cancerous and healthy tissues, including fibro, fat, and fibrous tissue.
Bowman et al. [27]	TPS Spectra 3000 System	Detected tumors at depths over 1 mm.
Grootendorst et al. [28]	TPI Handheld Probe System (TeraView Ltd., Cambridge, UK)	Discriminated between cancerous and healthy tissues with high accuracy.
Cassar et al. [29]	TPI and Spectroscopy	Achieved a cancer identification accuracy of 80%.
Bao et al. [30]	TeraPulse 4000 System (TeraView Ltd., Cambridge, UK)	Reached a spatial resolution of 1 mm.

Study	THz system	Key findings
Vohra et al. [31]	TPI System with Reflection Mode (TeraView Ltd., Cambridge, UK)	Cancerous areas exhibited the highest reflection, with THz results agreeing with histopathological analysis.
Okada et al. [32]	A Scanning Laser THz Near-Field Reflection Imaging System	Achieved a spatial resolution of 20 μm.
Bowman et al. [33]	TPS Spectra 3000 Pulsed THz Imaging and Spectroscopy System (TeraView Ltd., Cambridge, UK)	Showed higher absorption coefficients and refractive indices in cancerous areas compared to normal tissues, with a spatial resolution of 200 μm.
Vohra et al. [34]	Single point pulsed THz transmission spectroscopy (reflection imaging)	Tomographic images clearly distinguished between cancerous and healthy tissues.
Boutaayamou et al. [35]	THz TDS (Tera Pulse 4000) (reflection mode)	THz image contrast was well correlated with histopathological results.
Kaurav et al. [36]	sub-terahertz waveguide iris probe	Able to detect fat, fibrous, and cancer tissues.
Cassar et al. [37]	Refractive index classification with morphological dilation	Achieved sensitivity rates of 80% and 82% for different tissue types.

Table 1.
Relevant THz breast imaging exploration.

conditions and prevent contamination. The tumor sample was carefully positioned on Whatman filter paper to remove excess culture medium and clear the tissue of any residual fluids or tumor secretions. It was periodically moved to a dry region of the filter paper as it became saturated while absorbing excess fluid and medium. The sample was allowed to dry for at least 5 minutes to ensure complete removal of moisture before proceeding with further analysis. The tumor sample is securely positioned between two quartz windows and screwed tightly. A customized sample holder has been designed to place the sample for the measurement. The device has been fully automated for raster scanning of the tissue sample. The measurements are carried out at lower THz frequency, from 0.12 to 0.16 THz with 1 GHz spectral resolution for rapid scanning of the tissue sample. Measurements are carried out at 300 and 100 ms integration time, which takes 35 mins and 15 mins, respectively.

Figures 7–9 present a few of our THz measurement results on freshly excised tissue samples, with each row showing (a) a visible image of the sample, (b) the corresponding THz intensity map, and (c) the pathology slide. The color-coded THz maps illustrate variations in intensity, with red indicating higher intensity cancerous

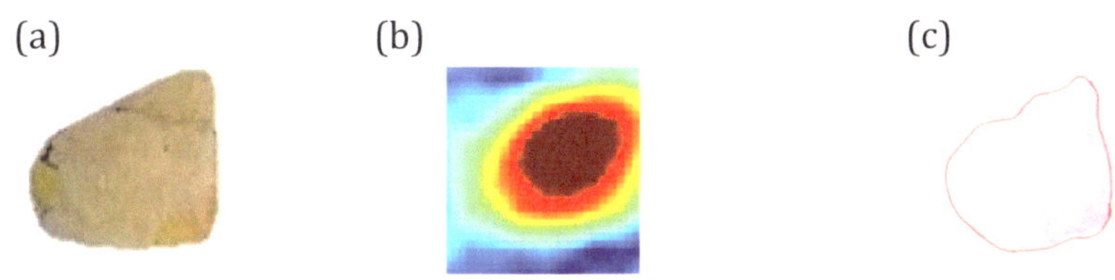

Figure 7.
(a) Visible image of breast tumor sample (b) THz image of the sample and (c) Histopathology image of the sample.

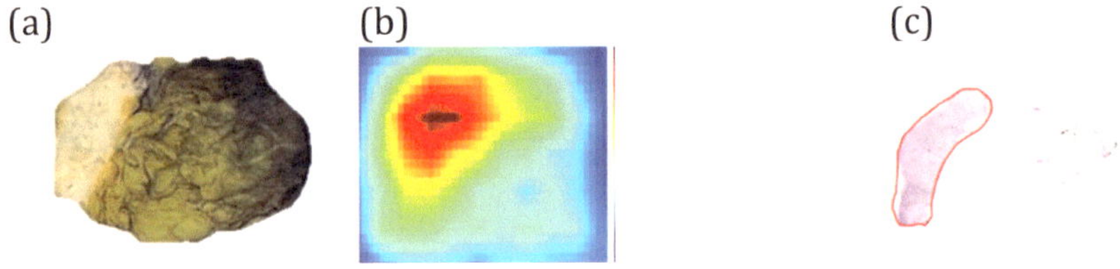

Figure 8.
(a) Visible image of breast tissue sample (b) THz image of the sample and (c) Histopathology image of the sample. The sample has both tumor and adjacent normal tissue. The tumor portion shows higher reflected intensity.

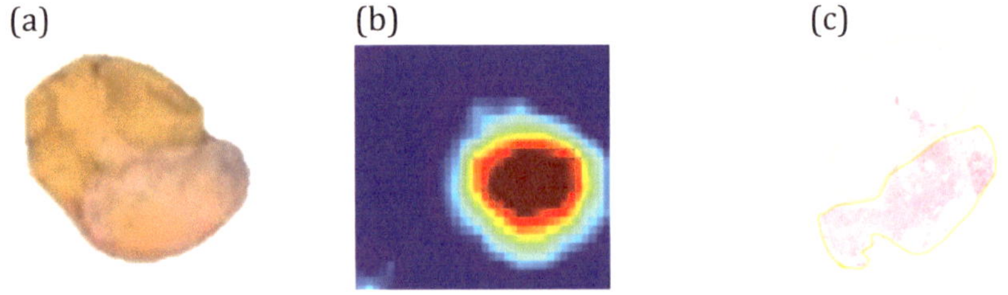

Figure 9.
(a) Visible image of breast tissue sample (b) THz image of the sample and (c) Histopathology image of the sample. The sample has both fibrous and stroma, in which the fibrous portion shows higher reflected intensity. The marking has been done by an expert pathologist.

regions and blue representing lower intensity non-cancerous regions. These maps are compared against the pathology slides to identify and delineate cancerous regions. **Figure 7** displays freshly excised breast tumor tissue samples, which are rightly picked up in the THz image. **Figure 8** shows a sample with adjacent normal tissue. The THz intensity map in **Figure 8** (**b**) accurately highlights cancerous regions in red, which correlates well with the pathology image (c). In **Figure 9**, a non-tumor sample is shown, where the THz intensity map confirms the high-intensity regions, due to the presence of fibrous tissue.

The table indicates that THz intensity at lower frequency can differentiate tumor and non-tumor (fat) tissue samples and the contrast in the fresh tissue sample is much better. However, the THz average intensity is also very high in the case of fibrous tissue and hyperplasia tissues as shown in **Figure 9** [33]. This is mainly due to the increase in cellular density in these tissue types compared to the surrounding normal fat tissues. Advances in THz systems, such as THz optical fiber imaging systems, THz imaging systems based on Schottky diodes, and time-of-flight estimation techniques, have improved resolution for better tissue distinction [38].

3.3 THz imaging and spectroscopy for oral cancer margin detection

Oral cancers present a different set of challenges due to the complex anatomy of the oral cavity and the need for precise detection of tumor margins to avoid disfigurement and ensure complete cancer removal. THz imaging and spectroscopy offer non-invasive techniques to address these challenges by providing detailed contrast between cancerous and healthy tissues based on their water content and other molecular characteristics. Studies on ex vivo oral cancer tissues have demonstrated the capability of THz imaging to differentiate between healthy and malignant tissues.

This is particularly important in oral cancer surgeries, where precise margin detection can significantly impact patient outcomes. Although in vivo studies are still limited, there is potential for THz technology to be integrated into clinical practice for real-time assessment during surgeries.

Yookyeong Carolyn Sim et al. explore the feasibility of using THz imaging for detecting oral cancer by analyzing excised tissues at frozen temperatures. They conducted THz imaging on seven oral cancer samples at both room temperature and −20°C, comparing the results to histological examinations. The findings demonstrated that THz imaging at frozen temperatures provided better sensitivity and contrast in distinguishing cancerous tissues from normal mucosa. Additionally, the study revealed that frozen THz imaging could detect tumors hidden below the tissue surface, which is challenging with conventional imaging methods. The improved contrast was attributed to differences in the cellular structure and water content between cancerous and healthy tissues, making THz imaging at frozen temperatures a promising technique for oral cancer detection [39]. It has also been shown that lower temperatures enhanced the contrast between malignant melanoma and healthy tissues [40].

Yang et al. introduced a novel label-free real-time high-resolution near-field THz spectroscopic imaging technique, specifically designed to enhance the intraoperative diagnosis of cancerous tissues, particularly focusing on different types of oral tissues such as cancerous tissues and normal epithelial tissue, connective tissue, and muscle tissue [41]. This method is noted for its ability to achieve real-time imaging rates as high as 20 frames per second at a resolution up to λ/10 at 1.0 THz in near-field conditions. It offers a significant improvement in spatial resolution compared to traditional THz time-domain spectroscopy. The imaging system can clearly distinguish between different tissue types based on their THz responses, offering the potential for accurate tumor margin detection during surgery. The system's real-time imaging capability is particularly advantageous for intraoperative settings, where time-sensitive decisions are critical. This method is faster than conventional THz imaging techniques, such as probe scanning THz microscopy, making it more suitable for real-time surgical applications. The study compared THz imaging results to traditional histopathological methods, demonstrating that the THz images could reliably identify the same regions of cancerous tissue. Also, the authors suggest that further enhancements, improving image quality, could broaden the applicability of this technique to other biomedical fields. Similar to breast tissue sample measurements, TeraLumen has also utilized its CW THz device, which is a portable compact system to be placed in a clinical environment. We have validated more than 110 tissue samples in this system.

Figures 10–12 present a few results of THz measurements on freshly excised oral tissue samples. **Figure 10** displays freshly excised oral tumor tissue samples, which are rightly picked up in the THz image.

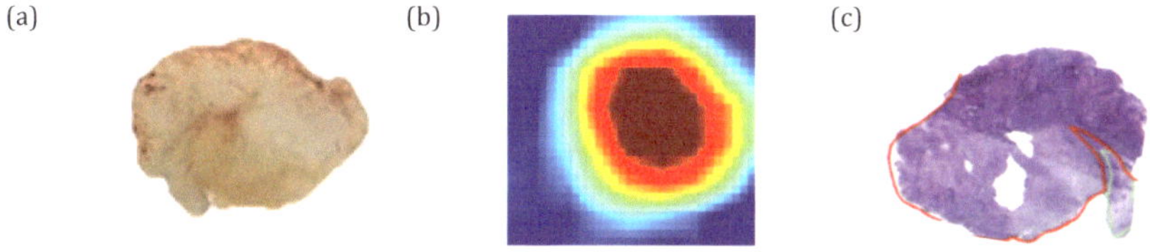

Figure 10.
(a) Visible image of oral tissue sample (b) THz image of the sample and (c) Histopathology image of the sample. The sample is completely tumor tissue, which has higher THz reflected intensity.

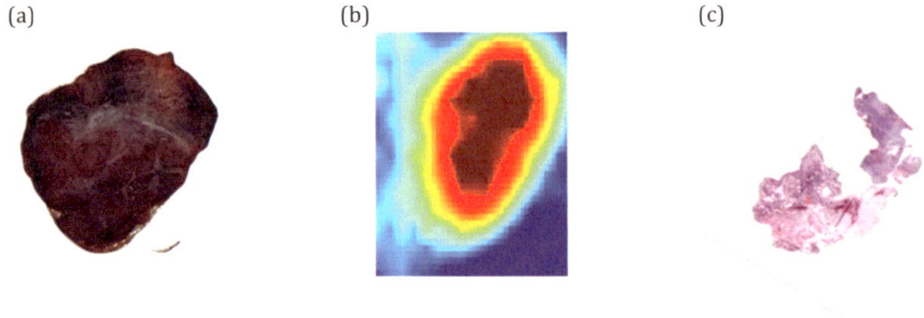

Figure 11.
(a) Visible image of a muscle tissue (b) THz image of the sample and (c) Histopathology image of the sample. THz reflected intensity shows higher amplitude, similar to tumor samples.

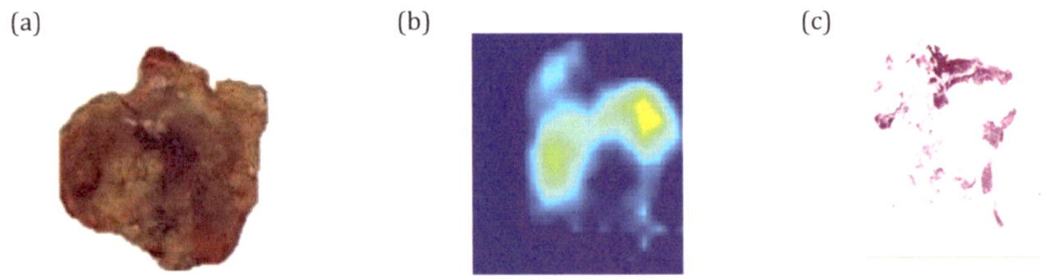

Figure 12.
(a) Visible image of oral normal mucosa tissue sample (b) THz image of the sample and (c) Histopathology image of the sample. The THz image correlates with the pathology image.

Figure 11 shows a sample with muscular tissue. The THz intensity map, **Figure 11** (**b**), shows higher reflected intensity in muscular tissue, due to its higher density. In **Figure 12**, a non-tumor sample, i.e. a mucosa, is shown, where the THz intensity map confirms the low-intensity regions. Advancements in image processing and deep learning algorithms will aid the inherent advantage of THz imaging in differentiating benign and malignant tissues.

4. Advanced AI algorithm

Enough research and studies have been carried out that leverage statistical interference to classify tissue regions from THz images. It has been observed that segmentation is challenging for images of fresh tissue and tumors that have invaded muscular regions [19]. Traditional imaging research has focused on qualitative analysis and statistical inference. The application of advanced data analytics algorithms, particularly deep learning, enables automatic quantification and optimized tissue analysis accuracy. Gezimati et al. have given a broad overview of recent usage of Artificial Intelligence (AI), particularly deep learning methods on THz sensing and imaging applications for medical detection [42]. Haoyan et al. have used wavelet synchronous-squeezed transformation (WSST) to convert the time-sequential THz data of each THz pixel to a spectrogram [19]. These spectrograms are used as an input to the deep neural network, which is used for the classification of the tissue. CNN-based classifiers have shown higher consistency between Terahertz scan and segmented results as shown in **Figures 13** and **14**.

This segmentation map in freshly excised tissue provides the primary evidence of the tissue segmentation, which in turn will help the physicians to take the clinical decision on margin detection. These segmentation algorithms help in precise

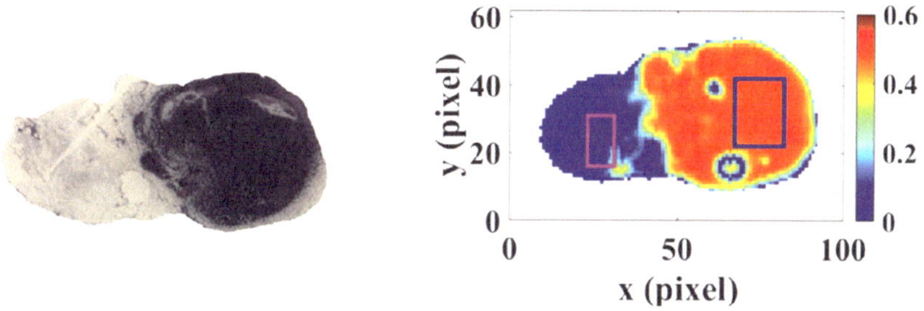

Figure 13.
Tumor sample with respective histopathology image and THz image along with selected region of training data (blue: cancer and red: fat training data) [19].

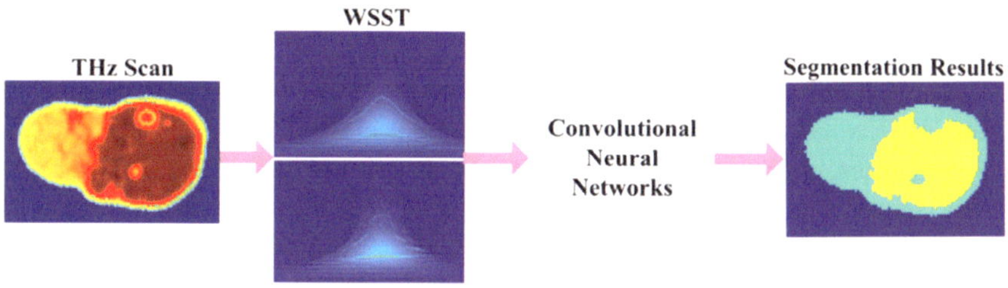

Figure 14.
Segmentation of Murine cancerous tissue with classic CNN neural networks [19].

boundary determination, crucial for detecting cancer tumor margins during surgeries. Developing computer-aided diagnosis (CAD) systems using machine learning and deep learning can enhance clinical decision support, addressing the limitations of manual methods, which are time-consuming and subject to interpretation variability.

AI in deep learning neural networks has made significant advancements in medical imaging over the last decade. Implementation of AI algorithms on THz imaging could potentially increase the region detection rates in freshly excised human tissue samples. Supervised machine learning methods such as Support Vector Machine (SVM), random forest (RF), and decision tree have been extensively utilized to perform THz-based segmentation and classification [19]. We have utilized both supervised learning SVM and RF as well as CNN to classify benign and malignant tissues in both breast and oral cancer. These models have been utilized to differentiate cancerous tissue at each pixel of the tissue image. CNN has achieved promising results for THz image segmentation due to the integration of spatial information. Hakeem et al. have also used THz Pulsed Imaging (TPI) to detect skin cancer, where they have applied image processing and Artificial Neural Networks (ANN). The ANN algorithm provides an accuracy of 94% [43]. The deep learning approach provides increased performance and automatic feature extraction in comparison to conventional machine learning techniques. This is very significant for the optimization of THz system design and further data analysis, which will, in turn, provide the roadmap for future research.

5. The promise and challenges of THz imaging in cancer detection

Along with various advantages, THz imaging also faces several challenges. The limited penetration depth of THz waves in biological tissues remains a significant hurdle, particularly for detecting cancers located deep within the body [44]. Additionally, the high cost and complexity of THz systems, especially pulsed systems, have hindered their widespread adoption in clinical settings. Despite these challenges, ongoing research and technological advancements continue to push the boundaries of what THz imaging can achieve, offering hope for its eventual integration into routine medical practice. As the field of THz imaging and spectroscopy continues to evolve, several critical challenges and promising opportunities have emerged, shaping the trajectory of future research and clinical applications. Addressing these challenges is essential to fully harness the potential of THz technology in medical diagnostics, particularly for cancer detection and treatment.

5.1 Scanning time improvement techniques

Another challenge is the long scanning times associated with THz imaging, caused by single-pixel detectors and the mechanical linear motion of optical delay lines. Using the motorized stages, imaging with better spatial resolution may take around 20 mins to half an hour of time for a single sample imaging. Techniques such as fast optical delay lines, cavity tuning, asynchronous optical sampling (ASOPS), electronically controlled optical sampling (ECOPS), and efficient data collection methods like 2D THz TDS and 2D electro-optic sampling (EOS) are being developed to accelerate scanning rates. The advent of near-field imaging, tomography, and technologies like 2D Galvano scanners and fast detector developments also contribute to reducing scanning times and achieving real-time imaging.

5.2 Edge scattering and pixel spread

Liu et al. have recently reported the potential challenges of the technology in the implementation of deep learning for breast cancer applications [19]. While comparing the THz images to the pathology images, the pixels near the edges have different spectral responses than any other region of the sample due to the scattering of the THz signal at the edges.

There are several regions along the boundary such as cancer and fat tissue, normal tissue, and background sample holder, where the tissues cannot be differentiated accurately. It is because of the spot size of the beam. Various point spread function (PSF) deconvolution algorithms such as Lucy deconvolution algorithms can be implemented to reduce the edge scattering as well as the beam spread.

5.3 Shape deformation between THz imaging and histopathology

THz images are mainly obtained in fresh tissue samples and the pathology images are obtained in the Formalin-fixed samples. Formalin fixing will dehydrate the sample and hence the shape used to vary significantly between both the images. Various image morphing algorithms have been utilized to match the size and shape of both images [45].

5.4 Technological advancements needed

THz technology has great potential to become a state-of-the-art medical imaging modality, complementing conventional techniques like X-ray and MRI. Its non-invasive, non-destructive, and non-ionizing nature, along with its ability to detect biomolecular rotational and vibrational energy levels, makes it suitable for label-free tissue differentiation. However, further advancements are necessary to address limitations such as high equipment costs, lack of standardized tools, high THz absorption by water, low resolution, low signal-to-noise ratio (SNR), and long scanning times. Overcoming these challenges will require developing high-speed THz imaging systems, real-time in vivo and non-invasive imaging configurations, and integration with modern technologies like robotics and the Internet of Things (IoT).

5.5 Future research directions

Future research should focus on enhancing diagnosis accuracy and analysis through the development of CAD systems, integrating THz technology with robotics, big data analytics, and cloud computing for remote patient monitoring. Investigating the impact of THz radiation on biological tissues, developing cost-effective, compact, and sensitive THz imaging systems for clinical trials, and establishing shared THz cancer image datasets will be crucial. TeraLumen has taken a significant step in building a compact and cost-effective device and carried out extensive clinical trials in the hospital environment. These advancements will support academic research and facilitate the clinical adoption of THz imaging technology.

6. Conclusion

THz imaging technology is emerging as a novel and innovative modality for medical diagnosis, particularly for cancer detection. It shows great potential to significantly enhance healthcare systems. The chapter discusses the application of THz technology in diagnosing breast and oral cancers. Additionally, the potential for cancer treatment through specific frequencies of THz radiation is highlighted. THz radiation's ability to differentiate tissues based on water content and its sensitivity to biomolecular rotational and vibrational energy levels further enhances its diagnostic capabilities. However, despite significant advancements, THz technology is still not mature enough for widespread clinical adoption.

Several challenges need to be addressed for THz imaging to become a mainstream medical imaging modality. These include high absorption of THz radiation by water limiting penetration depth. Techniques such as using thin or fixed tissue samples, proper geometry, and contrast agents like nanoparticles and glycerol can help improve penetration depth and image contrast. Also, THz imaging is characterized by long raster scan times. Techniques like fast optical delay lines, asynchronous optical sampling (ASOPS), electronically controlled optical sampling (ECOPS), and 2D electro-optic sampling (EOS) can help reduce scanning times and achieve real-time imaging. Moreover, advanced data analytic algorithms, particularly deep learning, can improve image quality and analysis. These algorithms enable precise boundary determination and the development of CAD systems for clinical decision support. Developing cost-effective, compact, and sensitive THz imaging systems and establishing shared THz cancer image datasets will support academic research and facilitate clinical trials. Though THz technology is a very good technology in differentiating cancer and normal tissue, distinguishing cancer and other tissue types such as dense fibrous tissues, muscles, and hyperplasia tissues is still challenging. Combining other multimodal techniques such as fluorescence imaging, Optical Coherence Tomography will help to differentiate all these tissue types. In addition, future development of handheld THz probe for in-situ applications will assist clinicians in making informed decisions during surgery.

Acknowledgements

We extend our sincere gratitude to Pfizer Indovation Grant, Menterra Social Impact fund, and BIRAC – DBT, NIDHI SSS fund for the financial support they have provided us. We would also like to thank our mentors Dr. Bala Pesala and Dr. Shyamsundar Mandyam for their guidance and extraordinary support on the product development. We would like to thank the doctors and supporting staff at Narayana Hrudayalaya and Mazumdar Shaw Medical center, Bangalore, India, for their clinical support.

Author details

Jyotirmayee Dash[1*], Sidharth Raj[1], Arun Jana[1], Vijay Pillai[2], Akshita Singh[3] and Nisheena Raghavan[4]

1 TeraLumen Solutions Private Limited, Chennai, Tamil Nadu, India

2 Department of Head and Neck Surgical Oncology, Narayana Health City, Bengaluru, India

3 Department of Breast Oncology and Oncoplastic Surgery, Narayana Health City, Bengaluru, India

4 Department of Histopathologist, Narayana Health City, Bengaluru, India

*Address all correspondence to: jyotirmayee.dash@teralumensolutions.com

IntechOpen

© 2024 The Author(s). Licensee IntechOpen. This chapter is distributed under the terms of the Creative Commons Attribution License (http://creativecommons.org/licenses/by/4.0), which permits unrestricted use, distribution, and reproduction in any medium, provided the original work is properly cited.

References

[1] O'Kelly Priddy CM, Forte VA, Lang JE. The importance of surgical margins in breast cancer. Journal of Surgical Oncology. 2016;**113**(3):256-263. DOI: 10.1002/jso.24047

[2] Arbyn M et al. Incomplete excision of cervical precancer as a predictor of treatment failure: A systematic review and meta-analysis. The Lancet Oncology. 2017;**18**(12):1665-1679. DOI: 10.1016/S1470-2045(17)30700-3

[3] Surci N et al. The rate of avoidable pancreatic resections at a high-volume center: An internal quality control and critical review. Journal of Clinical Medicine. 2023;**12**(4):1625. DOI: 10.3390/jcm12041625

[4] de Kleijn BJ et al. Intraoperative imaging techniques to improve surgical resection margins of oropharyngeal squamous cell cancer: A comprehensive review of current literature. Cancers (Basel). 2023;**15**(3):896. DOI: 10.3390/cancers15030896

[5] van der Pol CB et al. Accuracy of the liver imaging reporting and data system in computed tomography and magnetic resonance image analysis of hepatocellular carcinoma or overall malignancy—A systematic review. Gastroenterology. 2019;**156**(4):976-986. DOI: 10.1053/j.gastro.2018.11.020

[6] Hanneman K et al. Value creation through artificial intelligence and cardiovascular imaging: A scientific statement from the American Heart Association. Circulation. 2024;**149**(6):e296-e311. DOI: 10.1161/CIR.0000000000001202

[7] Movsisyan M, Al-Rossais AA, Sayeed S, Movsisyan G. Applications of terahertz waves in medical diagnostics: A literature review. International Journal of Community Medicine and Public Health. 2024;**11**(6):2450-2454. DOI: 10.18203/2394-6040.ijcmph20241512

[8] Koul SK, Kaurav P. Terahertz Spectrum in Biomedical Engineering. Singapore: Springer; 2022. pp. 1-29. DOI: 10.1007/978-981-19-3140-6_1

[9] Castro-Camus E, Koch M, Mittleman DM. Recent advances in terahertz imaging: 1999 to 2021. Applied Physics B. 2022;**128**(1):12. DOI: 10.1007/s00340-021-07732-4

[10] Stantchev RI, Yu X, Blu T, Pickwell-MacPherson E. Real-time terahertz imaging with a single-pixel detector. Nature Communications. 2020;**11**(1):2535. DOI: 10.1038/s41467-020-16370-x

[11] El-Shenawee M, Vohra N, Bowman T, Bailey K. Cancer detection in excised breast tumors using terahertz imaging and spectroscopy. Biomedical Spectroscopy and Imaging. 2019;**8**(1-2):1-9. DOI: 10.3233/BSI-190187

[12] Amini T, Jahangiri F, Ameri Z, Hemmatian MA. A review of feasible applications of THz waves in medical diagnostics and treatments. Journal of Lasers in Medical Sciences. 2021;**12**(1):e92. DOI: 10.34172/JLMS.2021.92

[13] Wang L. Terahertz imaging for breast cancer detection. MDPI. Sensors (Basel). 2021. DOI: 10.3390/s21196465

[14] Dash J, Ray S, Soniya S, Lenin B, Mandyam S, Pesala B. Development of compact, indigenous terahertz systems for medical diagnostics and NDT

applications. In: 2022 Workshop on Recent Advances in Photonics (WRAP). India: WRAP, IEEE; 2022. pp. 1-2. DOI: 10.1109/WRAP54064.2022.9758364

[15] Dash J, Jana A, Mandayam S, Pesala B, Dash SK. Detection of oral and breast cancer fresh tissue using continuous wave terahertz reflection imaging. In: Sadwick LP, Yang T, editors. Terahertz RF, Millimeter, and Submillimeter-Wave Technology and Applications XVII. SPIE OPTO; 2024. p. 63. DOI: 10.1117/12.3003861

[16] D'Arco A, Di Fabrizio M, Dolci V, Petrarca M, Lupi S. THz pulsed imaging in biomedical applications. Condensed Matter. 2020;**5**(2):25. DOI: 10.3390/condmat5020025

[17] Ghann W, Uddin J. Terahertz (THz) spectroscopy: A cutting-edge technology. In: Terahertz Spectroscopy - a Cutting Edge Technology. London, UK: InTech; 2017. DOI: 10.5772/67031

[18] Ashworth PC et al. Terahertz pulsed spectroscopy of freshly excised human breast cancer. Optics Express. 2009;**17**(15):12444. DOI: 10.1364/OE.17.012444

[19] Liu H, Vohra N, Bailey K, El-Shenawee M, Nelson AH. Deep learning classification of breast cancer tissue from terahertz imaging through wavelet synchro-squeezed transformation and transfer learning. Journal of Infrared, Millimeter, and Terahertz Waves. 2022;**43**(1-2):48-70. DOI: 10.1007/s10762-021-00839-x

[20] Wallace VP et al. Terahertz pulsed imaging of basal cell carcinoma ex vivo and in vivo. British Journal of Dermatology. 2004;**151**(2):424-432. DOI: 10.1111/j.1365-2133.2004.06129.x

[21] Fitzgerald AJ et al. Terahertz pulsed imaging of human breast tumors. Radiology. 2006;**239**(2):533-540. DOI: 10.1148/radiol.2392041315

[22] Chen H et al. Performance of THz fiber-scanning near-field microscopy to diagnose breast tumors. Optics Express. 2011;**19**(20):19523. DOI: 10.1364/OE.19.019523

[23] Joseph CS, Yaroslavsky AN, Neel VA, Goyette TM, Giles RH. Continuous wave terahertz transmission imaging of nonmelanoma skin cancers. Lasers in Surgery and Medicine. 2011;**43**(6):457-462. DOI: 10.1002/lsm.21078

[24] St. Peter B et al. Development and testing of a single frequency terahertz imaging system for breast cancer detection. IEEE Journal of Biomedical and Health Informatics. 2013;**17**(4):785-797. DOI: 10.1109/JBHI.2013.2267351

[25] Bowman TC, El-Shenawee M, Campbell LK. Terahertz imaging of excised breast tumor tissue on paraffin sections. IEEE Transactions on Antennas and Propagation. 2015;**63**(5):2088-2097. DOI: 10.1109/TAP.2015.2406893

[26] Bowman T, El-Shenawee M, Campbell LK. Terahertz transmission vs reflection imaging and model-based characterization for excised breast carcinomas. Biomedical Optics Express. 2016;**7**(9):3756. DOI: 10.1364/BOE.7.003756

[27] Bowman T, Wu Y, Gauch J, Campbell LK, El-Shenawee M. Terahertz imaging of three-dimensional dehydrated breast cancer tumors. Journal of Infrared, Millimeter, and Terahertz Waves. 2017;**38**(6):766-786. DOI: 10.1007/s10762-017-0377-y

[28] Grootendorst MR et al. Use of a handheld terahertz pulsed imaging device to differentiate benign and malignant breast tissue. Biomedical

Optics Express. 2017;**8**(6):2932. DOI: 10.1364/BOE.8.002932

[29] Bowman T et al. Pulsed terahertz imaging of breast cancer in freshly excised murine tumors. Journal of Biomedical Optics. 2018;**23**(02):1. DOI: 10.1117/1.JBO.23.2.026004

[30] Truong BCQ, Fitzgerald AJ, Fan S, Wallace VP. Concentration analysis of breast tissue phantoms with terahertz spectroscopy. Biomedical Optics Express. 2018;**9**(3):1334. DOI: 10.1364/BOE.9.001334

[31] Vohra N, Bowman T, Diaz PM, Rajaram N, Bailey K, El-Shenawee M. Pulsed terahertz reflection imaging of tumors in a spontaneous model of breast cancer. Biomedical Physics & Engineering Express. 2018;**4**(6):065025. DOI: 10.1088/2057-1976/aae699

[32] Okada K et al. Scanning laser terahertz near-field reflection imaging system. Applied Physics Express. 2019;**12**(12):122005. DOI: 10.7567/1882-0786/ab4ddf

[33] Bowman T, Vohra N, Bailey K, El-Shenawee M. Terahertz tomographic imaging of freshly excised human breast tissues. Journal of Medical Imaging (Bellingham). 2019;**6**(02):1. DOI: 10.1117/1.JMI.6.2.023501

[34] Vohra N, Bowman T, El-Shenawee M, Bailey K. Image reconstruction of freshly excised human breast tumors using terahertz electrical properties. In: 2019 IEEE International Symposium on Antennas and Propagation and USNC-URSI Radio Science Meeting. Atlanta, GA, USA: IEEE; 2019. pp. 571-572. DOI: 10.1109/APUSNCURSINRSM.2019.8888949

[35] Cerica D, Nguyen D, Hernandez Y, Verly JG, Boutaayamou M. Terahertz pulsed imaging of paraffin-embedded human breast cancer tissue. In: 2019 44th International Conference on Infrared, Millimeter, and Terahertz Waves (IRMMW-THz). Paris, France: IEEE; 2019. pp. 1-2. DOI: 10.1109/IRMMW-THz.2019.8874525

[36] Kaurav P, Koul S, Basu A. Design of sub-terahertz waveguide iris probe for breast cancer tumor margin assessment. In: 2021 IEEE MTT-S International Microwave and RF Conference (IMARC). Kanpur, India: IEEE; Dec. 2019. pp. 1-3. DOI: 10.1109/IMaRC49196.2021.9714693

[37] Cassar Q et al. Association of the terahertz refractive index and morphological dilation operations for breast carcinoma detection. In: 2021 46th International Conference on Infrared, Millimeter and Terahertz Waves (IRMMW-THz). IEEE; 2021. pp. 1-1. DOI: 10.1109/IRMMW-THz50926.2021.9567087

[38] Zhan X, Liu Y, Chen Z, Luo J, Yang S, Yang X. Revolutionary approaches for cancer diagnosis by terahertz-based spectroscopy and imaging. Talanta. 2023;**259**:124483. DOI: 10.1016/j.talanta.2023.124483

[39] Sim YC, Park JY, Ahn K-M, Park C, Son J-H. Terahertz imaging of excised oral cancer at frozen temperature. Biomedical Optics Express. 2013;**4**(8):1413. DOI: 10.1364/BOE.4.001413

[40] Sim YC, Ahn K-M, Park JY, Park C-S, Son J-H. Temperature-dependent terahertz imaging of excised oral malignant melanoma. IEEE Transactions on Terahertz Science and Technology. 2013;**3**(4):368-373. DOI: 10.1109/TTHZ.2013.2267415

[41] Yang S et al. Studying Oral tissue via real-time high-resolution terahertz

spectroscopic imaging. Physical Review Applied. 2023;**19**(3):034033. DOI: 10.1103/PhysRevApplied.19.034033

[42] Gezimati M, Singh G. Terahertz data extraction and analysis based on deep learning techniques for emerging applications. IEEE Access. 2024;**12**:21174-21198. DOI: 10.1109/ACCESS.2024.3360930

[43] Isam Hakeem S, Abdullah Hassoun Z. Skin cancer detection based on terahertz images by using Gabor filter and artificial neural network. IOP Conference Series: Materials Science and Engineering. 2020;**928**:032025. DOI: 10.1088/1757-899X/928/3/032025

[44] Gezimati M, Singh G. Terahertz imaging technology for localization of cancer tumours: A technical review. Multimedia Tools and Applications. 2023;**83**(11):33675-33711. DOI: 10.1007/s11042-023-16596-z

[45] Chavez T, Bowman T, Wu J, Bailey K, El-Shenawee M. Assessment of terahertz imaging for excised breast cancer tumors with image morphing. Journal of Infrared, Millimeter, and Terahertz Waves. 2018;**39**(12):1283-1302. DOI: 10.1007/s10762-018-0529-8

Section 4

THz Application Prospects

Chapter 6

Study of BL Lac Objects in Infrared and Submillimeter Region, Comparison with Radio Galaxies: Unification

Parag Seal

Abstract

Here, BL Lacertae objects, Fanaroff-Riley type I and type II radio galaxies, are studied from the optical region to the submillimeter region (0.44 to 850 μm)/(680–0.4 THz). Using the flux densities from optical (B/V) to Submillimeter Common-user Bolometer Array (SCUBA) observations, the spectral energy distributions (SED) are constructed and compared with radiative transfer models. Compared with radiative transfer models, it can be suggested that, in the mid-infrared and far infrared regions, the dust is ~50–100 K and in the near-infrared region ~1200 K. Submillimeter emission is mostly non-thermal synchrotron and contamination from the radio region, but there is also a contribution from thermal dust emission. The dust masses and cold dust masses of BL Lac objects and radio galaxies are determined. The emission mechanism depends mainly on the efficient or inefficient accretion flow, and there are other factors also. Comparing these characteristics of BL Lac objects with FR radio galaxies, it can be suggested that radio-selected BL Lacs are comparable to FR II radio galaxies, and X-ray-selected BL Lacs show more similarities with FR I radio galaxies and support unification schemes.

Keywords: BL lac objects, FR radio galaxies, infrared, submillimeter, dust masses, unification

1. Introduction

BL Lacertae objects are the most extreme class of extragalactic objects: radio-loud active galactic nuclei. Most of them show broadband continuum emission at all observed frequencies with almost no emission or absorption lines. This feature is very different from the spectral energy distribution of most other astronomical sources.

BL Lac objects have been discovered in radio and X-ray frequencies, and they have very strong, highly polarized and variable power law continua. Now there are more than 500 objects confirmed as BL Lac objects and are listed in the catalogs [1, 2]. At first, BL Lac objects are sorted mainly into two classes according to their broadband spectral properties: (1) "red" BL Lacs technically they are defined as LBLs (low-energy peaked BL Lacs) or RBLs/radio selected BL Lacs, the spectral energy distribution of most RBLs show peak frequency in the radio/infrared band and (2) "blue" or HBLs (high-energy peaked BL Lacs) or XBLs/X-ray selected BL Lacs, these XBLs show the peak in the UV/X-ray band. Later, an intermediate group of BL Lac objects (IBLs) is found, which shows the peak frequency in the optical region. The flux densities, luminosities and spectral indices α_{FIR} versus α_{NIR} show that these IBL samples bridge the gap between the subclasses RBLs and XBLs. These new large samples of BL Lac objects became available from the ROSAT All-Sky Survey – Green Bank (RGB samples) Catalog, which exhibits properties intermediate between the two classes RBLs and XBLs [3].

The BL Lac objects are extensively studied in the X-ray region [4–6], in the optical region ([7, 8] and references therein), and also in the radio region [9, 10]. 3CRR nearby galaxies are studied by Quillen et al. [11] in the submillimeter (870 μm) region and also compared the SEDs with BL Lac-type objects. López-Carniego et al. [12] studied blazers using the data provided by the Herschel-Astrophysical Terahertz Large Area Survey (H-ATLAS). Now, there are observations in the gamma-ray region with the Compton gamma-ray observatory (CGRO) and the Crimean astrophysical observatory [13, 14].

It is suggested that the low-energy peak in the spectrum is produced by synchrotron mechanism, and the high-energy peak is caused by the comptonization of the synchrotron radiation by the same population of electrons that produce the synchrotron component, synchrotron self-Compton (SSC) model [15, 16]. Giommi et al. [17] classified the BL Lac objects according to their peak energy of synchrotron power (γS peak). Low synchrotron peak (LSP) objects are mostly RBLs (peak ~3 μm < λ < 300 μm) and high synchrotron peak (HSP) objects are mostly XBLs (λ < 0.3 μm). The IBLs are intermediate synchrotron peak (ISP) objects, showing a peak at about 0.3 μm < λ < 3 μm.

There are different radiation models for relativistic jets to explain the emission mechanism. The beaming model where it is considered that XBLs are seen at larger angles to the line of sight than RBLs [1, 18]. But Sambruna et al. [10] and Fossati et al. [19] suggested that the difference in the spectra of RBLs and XBLs cannot be explained only by the beaming model. Georganopoulos and Marscher [20] suggested a model considering the orientation of the relativistic jet and the electron kinetic luminosity Λ kin of the jet, and that explains the difference in radiation mechanism in RBLs and XBLs more properly.

BL Lac objects are studied in the infrared region by 2MASS (2 Micron all-sky survey), ISO (Infrared Space Observatory), IRAS (Infrared Astronomical Satellite) and SPIRE (Spectral and Photometric Imaging Receiver [21]) in the near-infrared (NIR) to far infrared (FIR) region. These flux densities from optical to submillimeter region are available from NASA Extragalactic Database (NED)[1]. The Submillimeter Common-user Bolometer Array (SCUBA) is used to measure the submillimeter fluxes of galaxies and radio sources. The flux densities of extragalactic radio sources are

[1] http://ned.ipac.caltech.edu/forms/photo.html.

available from SCUBA observation and are calibrated, analyzed and explained in detail by Jenness et al. [22]. These space telescopes give a golden opportunity to study extragalactic objects in the FIR/submillimeter region and a great deal of information can be collected, mainly about dust emission.

To understand the emission mechanism properly, it is necessary to study the whole electromagnetic radiation range. Now, with the observation from gamma-ray to radio region flux densities, it is possible to have a better understanding of the emission mechanism of the SEDs and a detailed way to study the unification scheme also over the cold dust and the thermal emission of the interstellar medium.

The unified scheme of BL Lac objects and FR radio galaxies (RGs) are studied by many astronomers. FR RGs are classified into two types depending on their surface brightness: FR I and FR II type. FR I RGs are edge-darkened RGs and FR II RGs are core-darkened RGs. From the previous studies, it is suggested that BL Lac objects are low luminosity edge-darkened FR I RGs seen at different viewing angles [23, 24]. Some astronomers suggested that the diffuse radio emission from some RBLs is more consistent with FR II RGs than FR I RGs [25, 26]. Landt and Bignall [27] suggested that one third of the BL Lac sample can be identified with relativistically beamed FR II RGs. Giommi et al. [17] suggested that RBLs are like high-excitation radio galaxies (HERGs) and XBLs are similar to low-excitation radio galaxies (LERGs).

2. Sample selection

Earlier optical photometry was done for 24 BL Lac objects using the Bradford Robotic telescope by Seal & Baruch [28]; these objects are also studied in FIR and submillimeter region by Seal-Braun [29, 30]. Among those BL Lac objects, only those objects are selected here, and 2MASS, ISO, IRAS and SCUBA flux densities are available from NED[1]. Along with these objects, some more BL Lac objects are selected from the catalog [2], for which the flux densities are available. For comparison, the radio galaxies are studied, five FR I RGs and six FR II RGs, for which the flux densities are available from NED[1]. These FR I and FR II RGs are selected from the sample of radio-loud elliptical galaxies compiled by Zirbel and Baum [31].

3. Data analysis

Flux densities of 15 RBLs, 4 XBLs and 4 IBLs are available from NED[1]. For comparison flux densities of five FR I RGs and six FR II RGs are also used. The redshifts (z) of the objects are also available from NED[1].

3.1 Positional accuracy

Mao [32] did position correction for 833 BL Lac objects from 2MASS PSC (2MASS Point Source Catalog) and listed in **Table 1** by Mao [32]. Abrahamyan et al. [33] did a cross-correlation of IRAS PSC/FSC (Point Source/Faint Source Catalog) and WISE (Wide-field Infrared Survey Explorer) catalogs and created a catalog[2].

[2] http://vizier.u-strasbg.fr/viz-bin/VizieR?-Source=II/338.

Object Name RBL	Red shift z	Optical V	2 MASS 2.2 μm	ISO 25 μm	IRAS 60 μm	ISO 100 μm	ISO 180 μm	SPIRE 500 μm	SCUBA 850 μm
PKS 0048–097	0.444	1.18 ± 0.06	3.37 ± 0.137	195 ± 48	…	218 ± 115	< 933	…	374.2 ± …
PKS 0215 + 015	1.715	1.45 ± …	7.6 ± 0.161	171 ± 21	131 ± 17.0	239.0 ± 46.0	…	474.2 ± 20.6	568.0 ± …
PKS 0235 + 164	0.94	0.238 ± 0.006	3.39 ± 0.106	75[b] ± 18	256 ± 26	163.75 ± …	…	515.0 ± 18.9	2697.2 ± …
PKS 0521–36	0.056	7.84 ± …	19.6 ± 0.524	83.32[b] ± 22.5	354 ± 51	520.9 ± 113.0	…	2909.3 ± 72.7	2973.9 ± …
PKS 0537–441	0.896	1.73 ± 0.14	10.1 ± 0.090	70 ± 48	418 ± 37.6	647.0 ± 162.0	< 933	. .	993.6 ± …
PKS 0735 + 178	0.424	1.55 ± …	9.40 ± 0.121	225 ± 48	266 ± 17.0	523.0 ± 167.0	< 1029	. . .	852.9 ± …
PKS 0851 + 202 OJ 287	0.306	1.48 ± 0.039	8.16 ± 0.296	284 ± 48	963.0 ± 9.0	1454 ± 29.0	…	1417.6 ± 34.3	747.0 ± …
B2 1308 + 326 OP 313	0.998	0.094 ± …	6.80 ± 0.10	271 ± 29.7	4 19.8 ± 0.01	576.0 ± 133	…	466.0 ± 18.0	1507.3 ± …
S4 1418 + 54 OQ 530	0.152	5.22 ± …	14.9 ± 0.445	108 ± …	212.0 ± 29.7	204.0 ± 181.0	337.0 ± 251.0	…	170.0 ± 15.0
PKS 1749 + 09	0.320	1.21 ± …	3.25 ± 0.084	134 ± 48	226.0 ± 17.0	346.0 ± 102.0	<4146	…	2705.0 ± …
S5 1749 + 701	0.77	0.996 ± …	2.25 ± 0.131	132 ± 48	< 64.0 ± …	247.0 ± 60.0	<1725	…	346.1 ± …
S5 1803 + 784	0.684	1.54 ± …	3.31 ± 0.547	162 ± 48	440.0 ± 17.0	567.0 ± 57.0	507.0 ± 300	…	1269.8 ± …
S4 1807 + 698 3C 371	0.051	10.0 ± …	18.2 ± 0.536	172 ± …	232.0 ± 17.0	425.0 ± 66.0	…	…	764.8 ± …
S4 1823 + 56 4C 56.27	0.663	0.597 ± …	1.6 ± 0.262	63 ± 48	< 429[c]	< 411 ± …	459.0 ± 416.0		1209.6 ± …
PKS 2251 + 15 3C 454.3	0.859	2.03 ± …	3.85 ± 0.135	115 ± 18	183.0 ± 17	< 190 ± …	…	17850.5 ± 328.3	2634.7 ± …
XBL									
S5 1133 + 704 MRK 180	0.046	3.23 ± …	22.1 ± 1.50	9.61 ± 0.03	245.0 ± 22.0	508.0 ± 68.0	< 123	298.3 ± 19.9	210.0 ± 50.0
PKS 1514–241 AP Lib	0.049	3.25 ± 0.28	33.7 ± 1.70	169 ± 38	209.9 ± 29	1181.0 ± 95.0	…	689.7 ± 25.2	1875.8 ± …
S4 1652 + 398 MRK 501	0.033	4.71[a] ± …	93.2 ± 2.49	82.0 ± 22	107.0 ± 9.0	475.0 ± 99.8	…	…	220 ± 50.0
PKS 2155–304	0.116	36.6 ± …	50.9 ± 1.25	96 ± 12.0	93.0 ± 17	172.0 ± 72	1235.0 ± …	…	…

Object Name RBL	Red shift z	Optical V	2 MASS 2.2 μm	ISO 25 μm	IRAS 60 μm	ISO 100 μm	ISO 180 μm	SPIRE 500 μm	SCUBA 850 μm
IBL									
PKS 0754 + 100 OI 090.4	0.266	0.956 ± ...	3.03 ± 0.136	235 ± 13.4	293.0 ± 17.0	370.0 ± 44.0	...	...	1141.8 ± ...
PKS0829 + 046 OJ 049	0.180	3.24 ± ...	15.2 ± 0.416	41.7 ± 0.73	427.0 ± 17.0	490.0 ± 46.0	...	...	936.3 ± ...
S4 1101 + 384 MRK 421	0.031	23.8 ± 1.1	63.2 ± 1.81	95.0 ± 19.0	181.0 ± 22.0	361.0 ± 68.0	...	...	210.0 ± 50.0
1H 1730 + 500 1 ZW 187	0.055	1.88 ± 0.2-	5.64 ± 0.526	...	< 64	<170	...	...	

[a]B flux density.
[b]12 μm flux density.
[c]65 μm flux density.

Table 1.
Optical to submillimetre flux densities in mJy of BL Lac object.

3.2 Flux densities

Galactic extinction corrections are done following NED[1] using Galactic Extinction Calculator. k-corrections are done following k-correction calculator service[3] and from **Table 2**, Wisotzki [34]. The k-corrections for submillimeter flux densities are done only for objects with z > 0.5, following Archibald et al. [35].

These corrected magnitudes are converted to flux densities where necessary. Since BL Lac objects are point-like sources so the flux values derived and cataloged in the General Catalog are considered as the total flux values of the sources. Padovani et al. [36] suggested that the host galaxy contribution of flux in the IR region is negligible; the maximum is ~1%. So, these flux densities are used to derive the flux ratios, luminosities, dust masses, etc., for BL Lac objects. For comparison flux densities, luminosities of Radio galaxies are also used. These flux densities and redshift of the sample objects are listed in **Tables 1** and **2** [37].

3.3 Luminosity

The luminosities (Lλ) are estimated from the following relation

$$L\lambda = 4\pi D^2 F\lambda(\mathrm{LO}) \tag{1}$$

D is the luminosity distance in Mpc ($D = (zc/H_0)(1 + 1/2\,z)$), z is the redshift and Fλ is the flux density in Jy. D is estimated considering Hubble constant H_0 = 75 km s^{-1} Mpc-1, deceleration parameter $q_0 = -0.55$, for density parameter $\Omega_m = 0.34$ and $\Omega_k = 0.68$.

The optical luminosities (L_O) and submillimeter luminosities (L_S) are calculated using Eq. (1), from the flux densities at 0.54 μm (555.9 THz) and 850 μm (0.35 THz), respectively.

3.4 Spectral indices

The flux densities at 180 μm (1.66 THz), 60 μm (4.99 THz), 12 μm (24.98 THz) and 2.2 μm (136 THz) from ISO, IRAS and 2MASS observations are used to estimate FIR (α_{FIR}) and NIR (α_{NIR}) spectral indices.

The spectral indices are derived using the relations below Chen et al. [38],

FIR spectral index is

$$\text{á}_{180-60} = \log\left(\mathrm{F}_{180}\mu m / \mathrm{F}_{60}\mu m\right) / 0.477 \tag{2}$$

NIR spectral index is

$$\text{á}_{12-2.2} = \log\left(\mathrm{F}_{12}\mu m / \mathrm{F}_{2.2}\mu m\right) / 0.716 \tag{3}$$

3.5 Dust mass and dust temperature of BL lac objects

The dust temperatures of the objects are derived from flux densities at 100 μm (~3 THz) and 60 μm (~ 5 THz), considering the emissivity index β = 1, following

[3] http://kcor.sai.msu.ru/UVtoNIR.html.

DOI: http://dx.doi.org/10.5772/intechopen.1008264

Object name	RedShift z	Optical V	2 MASS 2.2 μm	ISO 25 μm	IRAS 60 μm	ISO 100 μm	ISO 180 μm	SPIRE 500 μm	SCUBA 850 μm
FR I									
S4 0316 + 41 3C 084	0.018	14.32 ± ...	80.2 ± 2.5	930 ± 100	7090 ± 750	8328.0 ± 64.0	4764.33 ± ...	2940.9 ± 79.7	2427.5 ± ...
PKS 0430 + 05 3C 120	0.033	7.61 ± ...	44.4 ± 1.1	289 ± 43.4	1383 ± 15	2786.0 ± 52.9	1383.9 ± ...	456.5 ± 25.4	1821.3 ± ...
PKS 0915–11 3C 218	0.0548	25.9 ± 5.49	13.6 ± 0.36	8.06 ± 0.029	155.0 ± 23.3	160.337 ± ...	160.45 ± ...	124.4 ± 10.8	69.3 ± ...
UGC 09799 3C 317	0.0344	21.9 ± 2.12	8.17 ± 0.136	4.12 ± 0.05	< 115	40.66 ± ...	47.90 ± ...	27.8 ± 14.2	91.0 ± 12.2
B2 1626 + 39 3C 338	0.030	39.2 ± 5.8	70.5 ± 10.4	8.80 ± 0.50	110 ± 24	139.356 ± ...	138.74 ± ...	...	73.0 ± 17.0
FR II									
PKS 0802 + 10 3C 191	1.956	0.141 ± ...	...	1.69 ± 0.16	...	...	21.9 ± 4.9	28.9 ± 8.1	6.39 ± 1.06
B2 1003 + 35 3C 236	0.099	1.86 ± 0.18	3.28 ± 0.139	16.7 ± 0.03	72 ± 10.8	93.0 ± ...	127.52 ± ...	72.7 ± 5.8	47.93 ± 10.14
B3 1206 + 439 3C 268.4	1.402	0.177 ± 0.004	0.776 ± 0.163	10.8 ± 0.07		74 ± 20	67.3 ± ...	9.9 ± 12.5	5.13 ± 1.18
1328 + 307 3C 286.0	0.85	0.458 ± ...	0.985 ± 0.132	7.64 ± 0.017	120 ± 40	126 ± 39	35.0 ± 26	112.7 ± 13.8	271.7 ± ...
Cygnus A 3C 405	0.057	3.67 ± 0.35	14.2 ± 0.317	250 ± 36	3034 ± 56	2155.2 ± 57.0	1256.44 ± ...	364.4 ± 114.9	560.0 ± 80.0
PKS 2313 + 03 3C 459	0.220	0.38 ± 0.037	1.69 ± 0.077	50.3 ± 0.04	683 ± 102	815.96 ± ...	604.14 ± ...	56.6 ± 11.8	...

Table 2.
Optical to submillimetre flux densities in mJy of radio galaxies.

Helou et al. [39]. Dust masses (M_d) can be estimated considering single dust temperature following Dunne et al. [40]:

$$M_d = F\lambda D^2 / k_d(\nu) B(\nu, T_d)(MO) \qquad (4)$$

Where $F\lambda$ is the flux density, D is the luminosity distance as explained in Eq. (1), the dust mass opacity coefficient κ_d is assumed as 0.077 m^2 Kg^{-1}, which is the intermediate value between graphite and silicon and $B(\nu, T_d)$ is the Planck function at the corresponding dust temperature T_d. The Planck functions are estimated from the distribution of β as a function of temperature T_d for galaxies following Dunne et al. [40].

4. Results

4.1 Flux densities and luminosities

The flux density at 850 μm (0.4 THz) versus the flux density at 160 μm (1.87 THz) is shown in **Figure 1**; This plot shows some scattered objects. The flux densities of FR I and FR II RGs are also shown for comparison. A linear regression line can be drawn and given as $\log F_{850} = 0.747 \log F_{160} - 1.62$. One FR II RG shows a very low submillimeter flux density compared to other RBLs and FR II RGs. Low submillimeter flux density indicates less cold dust and more dust at higher temperatures. Submillimeter fluxes of RBLs are higher than those of XBLs and IBLs but comparable to those of FR I and FR II RGs.

The optical luminosities (L_O) and submillimeter luminosities (L_S) are calculated using eq. 1, from the flux densities at 0.44 and 850 μm, respectively. The variation of submillimeter luminosity ($\log L_S$) with optical luminosity ($\log L_O$) is shown in

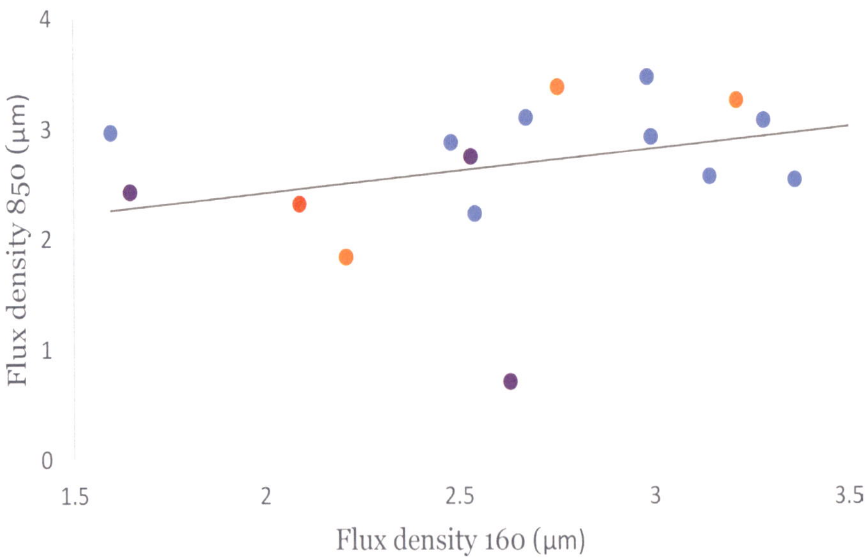

Figure 1.
Flux density at 160 μm vs. flux density at 850 μm. Blue circles RBLs, red circles XBLs, green circles IBLs, orange circles FR I radio galaxies, magenta circles FR II RGs. One regression line is shown.

Figure 2. For comparison luminosities of RGs are also shown in this figure. This variation of submillimeter luminosities with optical luminosities shows a good correlation. A linear regression line can be drawn and given as: log L_S = 1.76 log L_o–1.364. The submillimeter and optical luminosities of FR II RGs are comparable to those of RBLs, but FR I RGs show comparatively low optical and submillimeter luminosities and are comparable to those of XBLs. The plot shows an overlap at low optical luminosities. These luminosities of XBLs are smaller than those of RBLs but for IBLs these luminosities can be comparable with those of RBLs.

4.2 Spectral energy distribution

The SEDs of about 23 BL Lac objects from optical to submillimeter region are presented here (**Figures 3–5**); for all the sources, except one IBL sample and one FR II RG sample, submillimeter flux density at 850 μm are available in NED[1]. For about 20% of these sources, the flux densities in the FIR region are considered as upper limits and those flux densities are used to construct the SEDs.

The SEDs for RBLs are shown in **Figure 3**. For the RBL samples, the SEDs are extended up to 850 μm. The SEDs of almost all RBL samples show an increase in flux densities from NIR to FIR region; in the submillimeter region, the SED is flat or decreased, except for about 30% of the sources. Almost all the RBL samples show a bump ~60–100 μm.

Figure 4 shows the SEDs of XBLs and **Figure 5** shows the SEDs of IBLs. For XBLs, the SEDs show an increase in IR emission NIR to FIR region, then the SEDs are almost flat in submillimeter, except for one XBL sample. The SEDs of the IBLs show a gradual increase in flux densities from the NIR to submillimeter region. Except for one XBL sample, the XBLs and IBLs show a bump at ~10 μm. Only one XBL sample shows an increase in flux densities ~60–100 μm region.

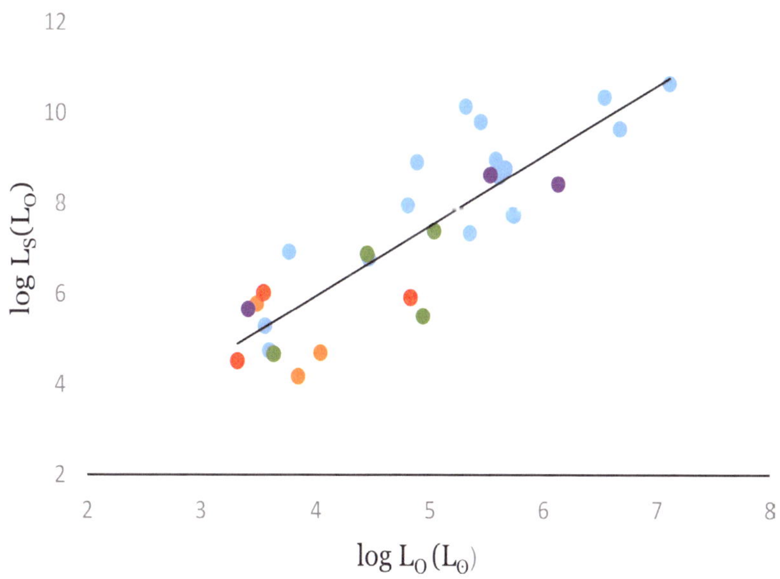

Figure 2.
The variation of submillimetre luminosity (log Ls (L⊙)) with optical luminosity (log LO (L⊙)). The symbols used are same as in ***Figure 1****. One regression line is shown.*

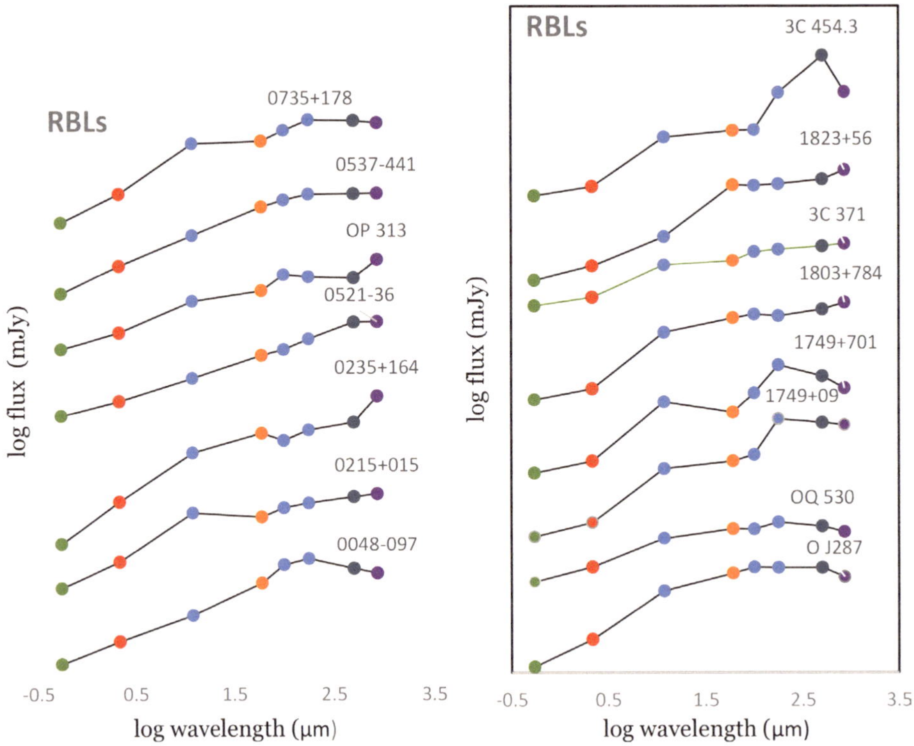

Figure 3.
Spectral energy distributions of RBLs. Green circle represents optical data and the average uncertainty is ±0.2, Red circle 2MASS data and average uncertainty is ±19.9, Orange circles IRAS data and average uncertainty is ±49.19, Blue circles ISO data and average uncertainty is ±85.8, Gray circles SPIRE data and average uncertainty is ±34.3, magenta circles SCUBA data and average uncertainty is ±15.5.

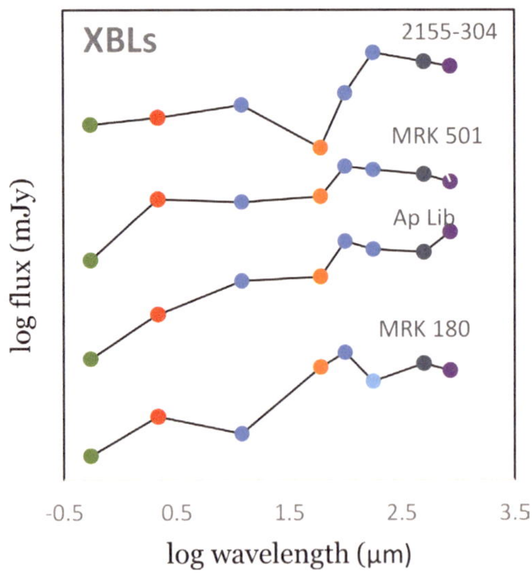

Figure 4.
Spectral energy distributions of XBLs. The symbols used are same as in ***Figure 3****. The average uncertainty for optical data is ±0.03, for 2MASS data is ±1.7, for IRAS data is ±44.7, for ISO data is ±41.5, for SPIRE data is ±22.5, for SCUBA data is ±28.5.*

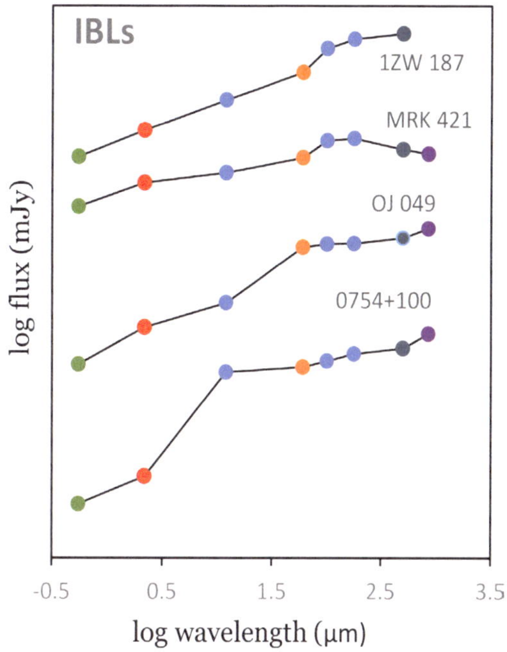

Figure 5.
Spectral energy distributions of IBLs. The symbols used are same as in ***Figure 3****. The average uncertainty for optical data is ±0.05, for 2MASS data is ±0.54, for IRAS data is ±40.5, for SPIRE data is ±30.3, for SCUBA data is ±18.5.*

Figures 6 and 7 show SEDs of FR I and FR II radio galaxies. About 60% of the FR I RG samples show a decrease in flux density in the submillimeter region. Except for two FR II RG samples, other FR II RGs show an increase in flux density in the submillimeter region.

Comparing the SEDs of all the sources with the radiative transfer models [41], **Figure 8** shows luminosities of BL Lac objects from MIR to FIR (12–100 μm) region are comparable L = $10^{9.0}$, $10^{9.5}$ L ʘ, radius R = 1 Kpc and visual extinction Av = 8 mag [41]. These luminosities are comparable to elliptical galaxies. For those sample sources where the flux densities remain almost the same or increase from ~60 to ~100 μm, luminosity ~ L = $10^{9.5}$ L ʘ, at radius R = 250 pc to 500 pc at visual extinction Av = 8 mag. The SEDs, in the NIR region, of all BL Lac objects and FR II type RGs, show luminosities L = 10^{11}, $10^{11.5}$ L ʘ at dust radius R = 1 Kpc with extinction Av = 2 mag. The SEDs of FR I RGs in the NIR region show luminosities L = 10^{12}, $10^{12.5}$ L ʘ, dust radius R = 1 Kpc and extinction Av = 2 mag.

4.3 Spectral indices

These spectral indices (α_{180-60}) versus ($\alpha_{12\text{-}k}$) of all the BL Lac samples and FR I and FR II RGs are plotted in **Figure 9**. This plot shows a big scatter. The FIR spectral indices (α_{180-60}) vary for RBLs α_{180-60} ~ −0.5 – 2.7, for XBLs α_{180-60} ~ −1 – 1.5, for IBLs α_{180-60} ~ 0.0–1.0. The NIR spectral indices ($\alpha_{12\text{-}k}$) vary for RBLs $\alpha_{12\text{-}k}$ ~ 0.5–2.5, for XBLs $\alpha_{12\text{-}k}$ ~ 0.5–1.0, for IBLs $\alpha_{12\text{-}k}$ ~ 0.0–0.5.

The FIR spectral indices (α_{180-60}) vary for FR I RGs α_{180-60} ~ 0.0–3.2, for FR II RGs α_{180-60} ~ −1.7 – 1.0. The NIR spectral indices vary for FR I RGs $\alpha_{12\text{-}k}$ ~ −1.0 – 1.0

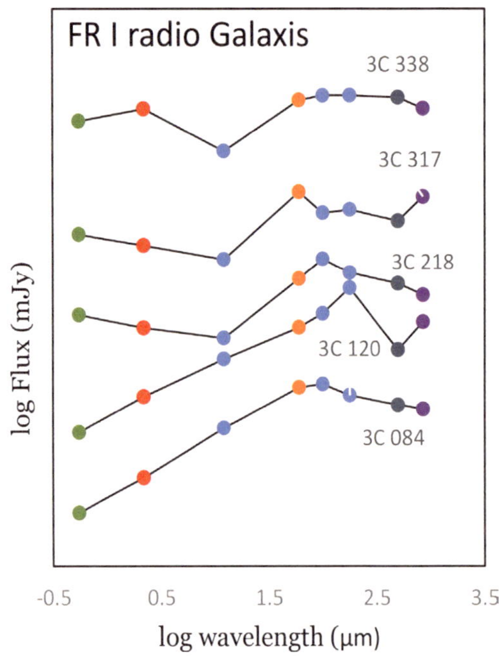

Figure 6.
Spectral energy distributions of FR I radio galaxies. The symbols used are same as in ***Figure 3****. The average uncertainty for optical data is ±0.17, for 2MASS data is ±19.8, for IRAS data is ±153.6, for ISO data is ±300.7, for SPIRE data is ±32.5, for SCUBA data is ±16.0.*

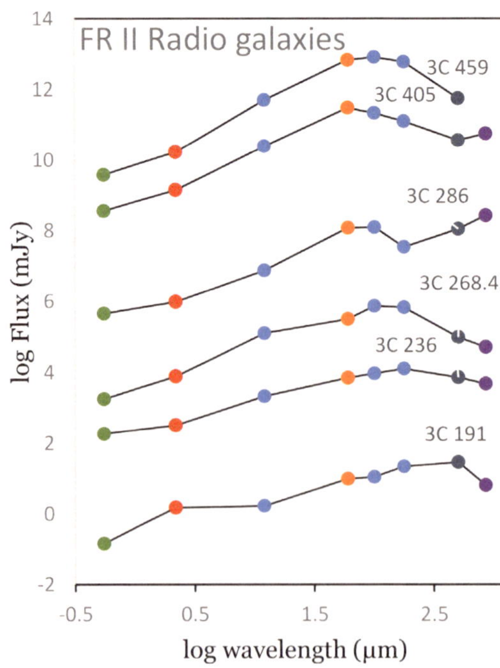

Figure 7.
Spectral energy distributions of FR II radio galaxies. The symbols used are same as in ***Figure 3****. The average uncertainty for optical data is ±0.10, for 2MASS data is ±0.025, for IRAS data is ±87.9, for ISO data is ±28.2. for SPIRE data is ±27.5, for SCUBA data is ±23.1.*

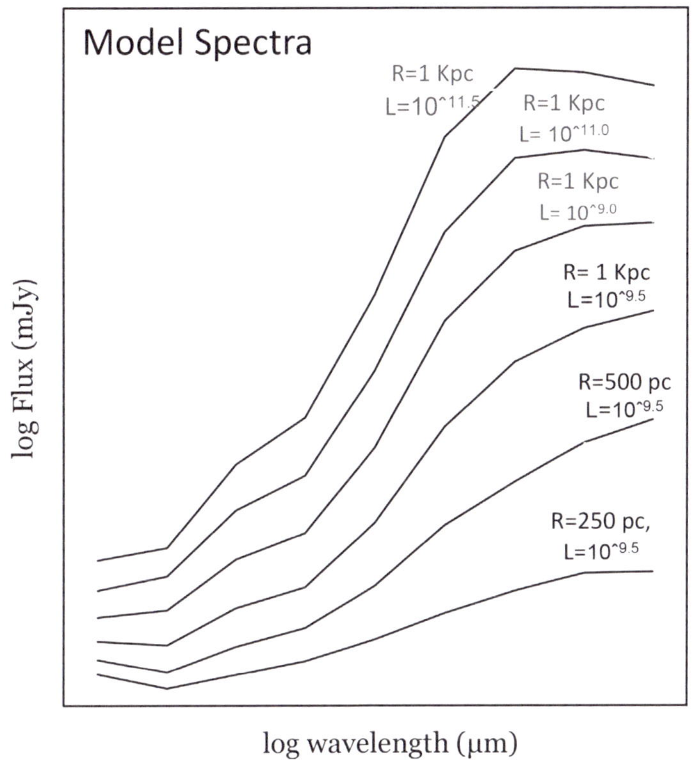

Figure 8.
Radiative transfer model spectra [41] for flux densities from 2.2 μm to 180 μm (136–1.66 THz), for visual extinction Av = 8 mag, radius R = 1 Kpc, Luminosity L = $10^{9.0}$ Lo and $10^{9.5}$ Lo; radius R = 250 pc and 500 pc, L = $10^{9.5}$ Lo; and Av = 2 mag, R = 1 Kpc, L = $10^{11.0}$, $10^{11.5}$ Lo.

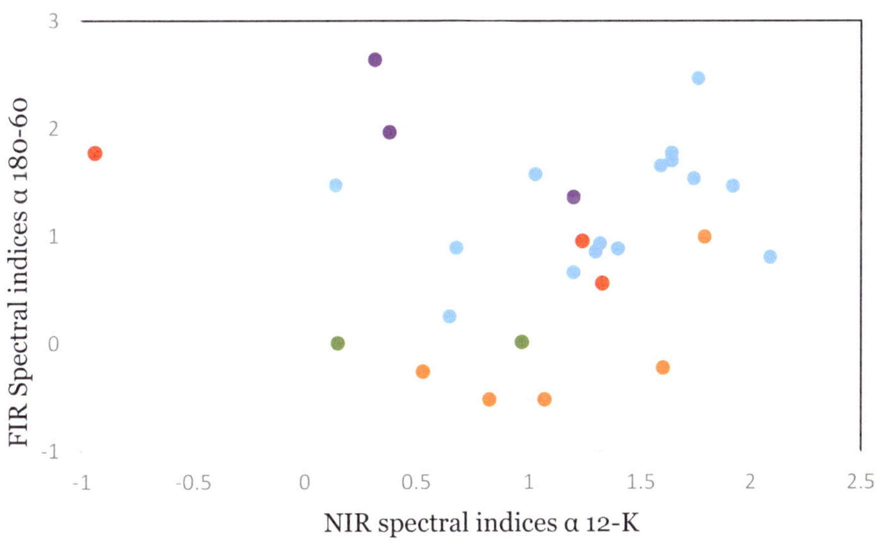

Figure 9.
The FIR spectral indices (α_{180-60}) vs. NIR (α_{12-k}) spectral indices. Blue circles are RBLs, red circles are XBLs, green circles are IBLs, orange circles are FR I radio galaxies, magenta circles are FR II radio galaxies. FIR average uncertainty ~ ± 69.3 mJy; NIR average uncertainty ~ ± 27.1 mJy.

and for FR II RGs $\alpha_{12\text{-}k} \sim 0.6\text{–}3.6$. The distribution of FIR and NIR spectral indices in **Figure 9** shows some overlap between RBLs and XBLs but not with radio galaxies. For FR I RGs, the NIR spectral indices are comparable to those of BL Lac objects, but FIR spectral indices are much lower, and for FR II RGs, the FIR and NIR spectral indices are comparable with RBLs. This is because of the different physical conditions of these sources.

4.4 Dust mass and dust temperature of BL lac objects

The dust masses (M_d) are estimated using flux density F_{100} at 100 μm and the cold dust masses (M_{cd}) are estimated using flux density F_{500} at 500 μm. The variation of logarithm of dust masses (M_d) with redshift (z) is shown in **Figure 10**. For RBLs, the dust masses remain almost the same order as the redshifts increase, but for XBLs and IBLs, the dust masses increase with an increase in redshift. The distribution in this plane shows an overlap at low redshift ($0.05 < z < 0.1$), the BL Lac objects with radio galaxies mainly FR I type RGs. BL Lac objects and FR II-type RGs have almost similar dust masses. For FR I radio galaxies, this variation shows a much steeper slope. In about 30% of FR I RG samples, the dust masses are comparable with XBLs; otherwise, the dust masses of FR I RGs are much lower. This is due to the low redshift of FR I RGs.

It is possible to calculate cold dust mass $M_{cd}(M\odot)$, considering the flux density at 500 μm (0.6 THz), considering cold dust temperature as 20 K and emissivity index $\beta = 2.0$, as suggested by Dunne et al. [40], using Eq. (4). The logarithm of (dust masses $M_d(M\odot)$ versus luminosities $L_{500}(L\odot)$ and cold dust masses $M_{cd}(M\odot)$ versus luminosities $L_{500}(L\odot)$) of BL Lac objects and a few radio galaxies are shown in **Figure 11**. This plot of the dust masses $M_d(M\odot)$ versus luminosities $L_{500}(L\odot)$ and cold dust masses $M_{cd}(M\odot)$ versus luminosities $L_{500}(L\odot)$ show a tight correlation. It is possible to draw a regression line and given as: $\log L_{500} = 1.01 \log M_{cd} + 0.191$.

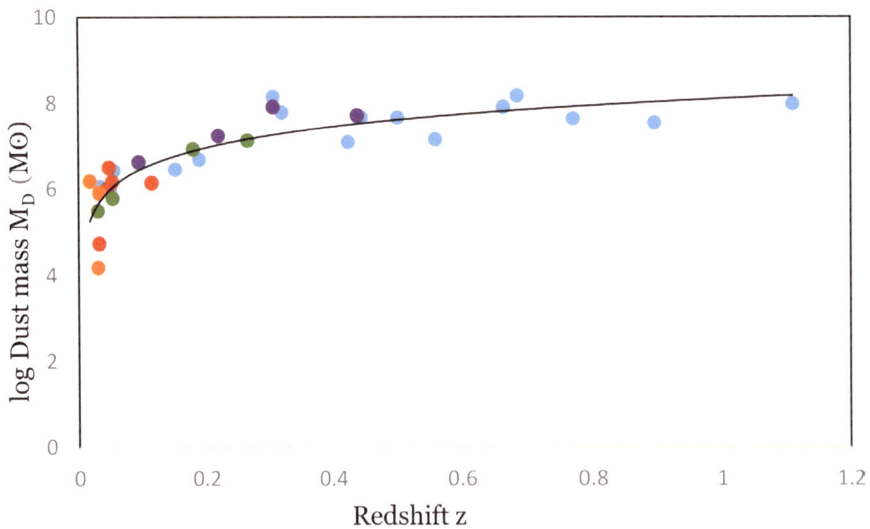

Figure 10.
*log Dust mass (log (Md (M⊙)), distribution against Redshift z. The symbols used are same as in **Figure 1**. One regression line is shown.*

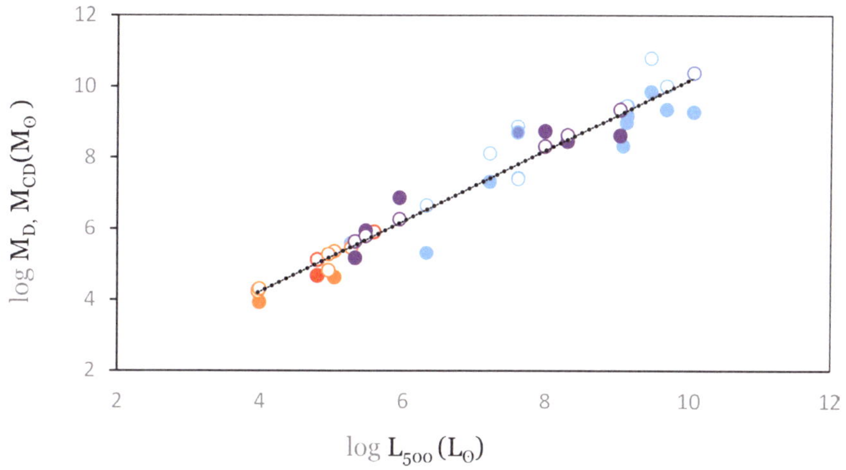

Figure 11.
*Luminosity log L_{500} (L⊙) vs. dust masses log (Md (M⊙), Mcd (M⊙)). The symbols used are same as in **Figure 1**. Filled circles are dust masses (Md (M⊙), open circles are cold dust masses (Mcd (M⊙). One regression line is shown.*

The cold dust masses of all the objects are slightly higher than dust masses. The cold dust masses of FR II RGs are comparable to those of RBLs and those of FR I RGs with those of XBLs.

5. Unification of BL lac objects and FR radio galaxies

The unifying model for low-power radio-loud active galactic nuclei predicts that BL Lacs and Fanaroff-Riley radio galaxies belong to the same class of objects, the difference being the orientation of their relativistic jets with respect to the line of sight.

From the previous studies, it is suggested that BL Lac objects are low luminosity edge-darkened FR I RGs; the difference is only in the viewing angles [23, 24]. Some astronomers, Cassaro [25] and Rector and Stocke [26], suggested that some RBLs are more consistent with FR II RGs than FR I RGs. Landt and Bignall [27] suggested that those BL Lac objects whose radio properties are comparable with FR I radio galaxies are weak-lined radio-loud AGNs and other BL Lac objects are comparable to FR II radio galaxies having similar radio properties as strong-lined radio-loud AGNs. So here, the flux densities, luminosities and dust masses of FR RGs are compared with those of BL Lac objects and are shown in the figures.

Figure 1 shows the flux densities of BL Lac objects and radio galaxies. The radio galaxies also follow the regression as the BL Lac objects, but there are a few scattered objects also. The flux densities of RBLs are comparable to those of FR II RGs, and the flux densities of XBLs are comparable to those of FR I RGs. Submillimeter luminosities and optical luminosities of BL Lac objects and radio galaxies are shown in **Figure 2**. Here also they show good correlation except for a few FR I RGs, which show low submillimeter and optical luminosities. There is a selection effect, the FR I RGs studied here are mostly low redshift objects. That is probably also the reason that the FR I RGs show less luminosity. But for almost 50% of FR I RGs the luminosities are comparable with those of XBLs. The luminosities of FR II RGs are comparable with those of RBLs. Dust masses versus redshift in **Figure 10** shows a good correlation.

Dust masses and cold dust masses versus luminosities of BL Lac objects and Radio galaxies are shown in **Figure 11**. It shows a very tight correlation and almost all the objects follow the regression line. The dust masses and luminosities of FR II RGs are comparable with those of RBLs and of FR I RGs with XBLs.

Since flux densities of large number of sample objects are not available in the literature, it is difficult to come to a conclusion. In all these results, we find the flux densities, luminosities and dust masses of FR II RGs are comparable to those of RBLs, and FR I RGs show more similarities with XBLs. So, this investigation supports the unifying scheme, and it indicates here that most RBLs show similarity with FR II RGs and XBLs with those of FR I RGs.

6. Discussion

Figure 1, flux densities at 850 μm versus 160 μm shows some scatter. According to the line of regression, those objects above the regression line indicate submillimeter excess. In the plot, a few RBL and FR I RG samples show high submillimeter and FIR flux densities and one FR II RG show less FIR and submillimeter flux density. There are different explanations for this submillimeter excess emission such as the variation of emissivity of the dust grains with temperature, dust grain size, spinning dust emission, and very cold dust, but it is still unclear. Lu et al. [42] informed about a core-jet structure with the jet extending to the northeast for RBL PKS 1749 + 096. Kino et al. [43] studied 3c 084 and suggested a strong collision between the jet and a compact dense cloud. The metal-poor galaxies and dwarf galaxies generally show submillimeter excess [44]. Leeuw et al. [45] suggested that the presence of diffuse dust heated by collision with electrons in the hot gas affects the determination of flux. Giommi et al. [17] reported that there is an accretion disk around FR II RGs and sources like these RGs and no accretion disk of FRI RGs and sources like FR I RGs. So, due to accretion flow, the fluxes of FR II RGs are higher than those of FR I RGs.

The variation of submillimeter luminosity (log L_s) with optical luminosity (log L_o) in **Figure 2** shows a good correlation except for a few scattered objects. The radiation emitted from BL Lac objects can be explained by the emission from plasma moving in the relativistic jet closely aligned to the line of sight. For XBLs and IBLs, the submillimeter and optical luminosities are lower than those of RBLs. For RBLs, the relativistic jets are oriented close to the line of sight at an angle of ~10°–15°, while that for XBLs is ~15°–30° [46]. So the observed flux is amplified if the relativistic jet angle is oriented closer to the line of sight than when the relativistic jet axis is not oriented close to the line of sight and radiation is projected. Willott et al. [47] studied radio galaxies and radio quasars and suggested that radio quasars are viewed with larger opening angles that allow a greater fraction of radiation to escape so the radio quasars are brighter than radio galaxies. Capetti and Celotti [48] suggested that the optical emission is also jet-related. Hardcastle and Worrall [49] studied the correlation between the luminosities in radio and the optical region of radio galaxies and argued that these components have a common origin at the base of the relativistic jet, which may also be true for BL Lac objects. Along with this, the nature of accretion disk also plays a key role. For XBLs, the inverse Compton emission dominates, so the thermal emission from inverse Compton on the external field of photons would modify the amount of soft X-ray emission [17]. All these suggestions give some explanation of the emission process from these sources.

The SEDs of BL Lac objects and also radio galaxies are shown in **Figures 3–5**, from optical region to submillimeter region. The SEDs of a few RBL samples, XBLs, IBLs and radio galaxies show bump at ~10 μm. Shi [50] concluded that the bump in the SEDs at ~10 μm is for excitation mechanism for aromatic features and is predominantly due to star formation. Radiative transfer model (**Figure 8**) also shows emission near ~11.3 μm is mainly due to PAH bands.

The SEDs of the RBLs show increase in IR emission at ~25 μm. Comparing with radiative transfer models, the emission at ~25 μm is due to large grains with temperatures ~300 K. The RBLs and radio galaxies show infrared excess in the MIR to FIR region (~ 60 μm) from hot dust at >100 K, either by large grains or small graphites.

The SEDs of about 50% RBLs, one XBL and radio galaxies studied here show peaks in infrared emission at ~100 μm and 180 μm region. This emission is mainly due to cold dust ~50 K. For excess infrared emission in the FIR region, the surrounding interstellar radiation field (ISRF) also plays a role.

For about 50% of RBLs and radio galaxies, the SEDs from MIR to FIR region are almost flat and continuous. Zang et al. [51] suggested that the synchrotron emission dominates from MIR to FIR region. Muller et al. [52] pointed out that for radio galaxies, the MIR and FIR infrared emission is mainly synchrotron radiation with negligible stellar contribution from any thermal source. For these objects, synchrotron emission dominates the MIR to FIR region. The infrared flux densities of BL Lac objects in the MIR to FIR region stay almost stable and the emission mainly synchrotron radiation. These objects have small dust at higher temperature, nearer to the nucleus and almost no PAH. As suggested in the literature these objects may be highly beamed objects where synchrotron emission dominates. Plotkin et al. [53] suggested that for highly beamed objects the synchrotron radiation dominates but for weakly beamed objects there is lack of observable torus emission and these BL Lac objects are fed mainly by radiatively inefficient accretion disks.

Figure 9 shows the FIR and NIR spectral indices FIR (α_{FIR}) versus NIR (α_{NIR}). The distribution of FIR and NIR spectral indices in **Figure 9** shows almost no overlap. The plot shows the FIR (α_{FIR}) spectral indices of RBLs and FR I RGs almost of the same order, whereas the FIR spectral indices of FR II RGs are negative, on average ~ −0.4. This indicates there may be more cold dust present in BL Lac objects and FR I RGs than in FR II RGs. The NIR spectral indices (α12-k) of BL Lac objects and FR II RGs are almost of the same order (~1.6), whereas those of FR I RGs (~ – 0.2) are different. In the NIR region, the physical properties of dust present in BL Lac objects and FR II RGs may be in a similar state.

Figure 10 shows the variation of the logarithm of dust masses for BL Lac objects and RGs with redshift. The logarithm of dust masses of RBLs and FR II RGs vary almost linearly with redshifts, whereas that variation for FR I radio galaxies is much faster and the slope is much steeper. The FR I RGs have relatively smaller dust masses than BL Lac objects and FR II RGs. The plot shows an overlap at a small redshift, z ~ 0.05. BL Lac objects, mainly RBLs and FR II RGs, have similar dust masses. The dust masses of XBLs are comparable to those of FR I RGs and dust masses of IBLs are comparable to those of low redshift RBLs. The FR I RGs selected here are very low redshift (0.01 < z < 0.15) objects. The variation of dust masses FR I radio galaxies shows a steep slope almost 75 times faster than BL Lac objects.

For BL Lac objects, $\log M_d = 1.03\,z + 6.68$ (25 K < T < 65 K).

For FR I radio galaxies, $\log M_d = 74.34\,z + 3.03$ (25 K < T < 50 K).

For FR II radio galaxies, $\log M_d = 1.04\,z + 6.77$ (25 K < T < 65 K).

The dust masses and cold dust masses of BL Lac objects and RGs versus luminosities are studied here (**Figure 11**). Cold dust masses are derived considering flux density at 500 μm. The cold dust masses are slightly higher than the dust masses estimated considering flux density at 100 μm. The derived dust masses provide a rough estimate of interstellar material associated with these objects.

Wang et al. [54] suggested that FR II RGs tend to have clumpy dust, and some of the BL Lac objects, mainly the RBLs have also clumpy dust as clumpy-ADAF (advection-dominated accretion flow) mode. The dust present in FR I RGs is not in clumpy form and possibly has not been detected. That explains the higher dust masses in RBLs and FR II RGs than in FR I RGs. But, the dust detected in clumpy structures does not correspond to all of the dust in the galaxies and may represent only a small (but variable) proportion.

For higher flux densities of RBLs, several astronomers suggested different theories. The RBLs show their energy peak at infrared/radio bands. As suggested by Giommi [17], the RBLs are like HERG and FR II RGs with the presence of accretion disks. So along with the jet emission, there is also emission from the accretion disk, and also thermal emission from dust. The excess submillimeter emission also indicates the presence of cold dust at about T ~ 10 K. For XBLs, the energy peak is at X-ray/UV region mainly due to the synchrotron self-Compton process but less thermal emission from dust. The relativistic jet power, viewing angle and the accretion disk play the actual role. The accretion disk emission, the jet emission and thermal emission make the submillimeter emission higher.

It is not possible to find a large number of sample objects in the catalogs, still comparing flux densities, luminosities, dust masses, etc., of these sample objects, it can be suggested that RBLs are like high-excitation radio galaxies (HERGs), FR II RGs and XBLs are similar to low-excitation radio galaxies (LERGs), FR I RGs and supports unification. It may be that FR II RGs are the parent population of RBLs and FR I RGs are the parent population of XBLs [23].

7. Conclusion

Here the flux densities, luminosities, dust masses, etc., for BL Lac objects: RBLs, XBLs, IBLs and FR type I and type II radio galaxies are studied from optical region to submillimeter region (0.44 to 850 μm) / (680–0.4 THz) to understand the dust emission.

The SEDs of these objects are compared with radiative transfer models, and from there, it can be concluded that, in the MIR and FIR region, the infrared emission is from the dust containing ~50–100 K and in the NIR region dust ~1200 K.

The flux densities, luminosities, etc., of RBLs are comparable with those of FR II RGs. This indicates that the RBLs are like high-excitation radio galaxies (HERGs) and FR II RGs with the presence of accretion disks. So along with the jet emission there is also emission from the accretion disk, and also thermal emission from dust.

The flux densities, luminosities, etc., of XBLs are comparable to those of FR I RGs and low-excitation radio galaxies (LERGs). For XBLs the emission is mainly due to synchrotron self-Compton process, but less thermal emission from dust.

Comparing these properties of BL Lac objects and radio galaxies, it can be suggested that RBLs are like HERGs, FR II RGs and XBLs are similar to LERGs, FR I RGs and it supports unification.

To study more accurately the emission and physical nature of these objects, it may be necessary to study a more detailed model from gamma-ray to radio emission, to get a more clearer view.

> *'The greater the sphere of our knowledge, the larger is the surface of its contact with the infinity of our ignorance' ...A Short History of Astronomy, A. Berry.*

Author details

Parag Seal
Independent Researcher, Groningen, The Netherlands

*Address all correspondence to: sealbraun@gmail.com

IntechOpen

© 2025 The Author(s). Licensee IntechOpen. This chapter is distributed under the terms of the Creative Commons Attribution License (http://creativecommons.org/licenses/by/4.0), which permits unrestricted use, distribution, and reproduction in any medium, provided the original work is properly cited.

References

[1] Padovani P, Giommi P. The connection between X-ray and radio-selected BL Lacertae objects. The Astrophysical Journal. 1995;**444**:567-581

[2] Veron-Cetty MP, Veron P. A catalogue of quasars and active nuclei: 13th edition. Astronomy and Astrophysics. 2010;**518**:A10-A18

[3] Laurent-muehleisn SA, Kollgaard RI, et al. Radio-loud active galaxies in the northern ROSAT all-sky survey. I. Radio identifications. A & A Supplement Series. 1997;**122**:235-248

[4] Böttcher M, Chiang J. X-ray spectral variability signatures of flares in BL lac objects. The Astrophysical Journal. 2002;**581**:127-165

[5] Rector TA, Stocke JT, Perlman ES, et al. The properties of the x-ray-selected emss sample of BL LAC objects. Astronomy Journal. 2000;**120**:1626-1658

[6] Padovani P, Costamante L, Giommi P, et al. BeppoSAX observations of 1-Jy BL Lacertae objects – II. Monthly Notices of the Royal Astronomical Society. 2004;**347**:1282-1295

[7] Nilson K, Pursimo T, Heidt J, et al. R-band imaging of the host galaxies of RGB BL Lacertae objects. Astronomy and Astrophysics. 2003;**400**:95-118

[8] Falomo R, Pian E, Treves A. An optical view of BL Lacertae objects. Astronomy and Astrophysics Review. 2014;**22**:73-117

[9] Giommi P, Ansari SG, Micol A. Radio and X-ray energy distribution of BL lac objects. Astronomy and Astrophysics Supplement Series. 1995;**109**:267-291

[10] Sambruna RM, Maraschi L, Urry CM. On the spectral energy distributions of blazars. The Astrophysical Journal. 1996;**463**:444-465

[11] Quillen AC, Almog J, Yukita M. 870 micron observations of nearby 3CRR radio galaxies. The Astronomical Journal. 2003;**126**:2677-2697

[12] López-Carniego M, González-Nuevo J, Massardi M, et al. Mining the Herschel★ -ATLAS: Submillimeter-selected blazars in equatorial fields. Monthly Notices of the Royal Astronomical Society. 2013;**430**:1566-1579

[13] Neshpor YI, Chalenko NN, Stepanian AA, et al. BL lac: A new ultrahigh-energy gamma-ray source. Astronomy Report. 2001;**45**:249-254

[14] Ackermann M, Ajello M, Atwood WB, et al. The third catalog of active galactic nuclei detected by the Fermi large area telescope. The Astrophysical Journal. 2015;**810**:14-98

[15] Ghisellini G, Padovani P, Celotti A, et al. Relativistic bulk motion in active galactic nuclei. The Astrophysical Journal. 1993;**407**:65-82

[16] Marscher AP, Gear WK. Models for high-frequency radio outbursts in, extragalactic sources with application to the early 1983 millimeter-to-infrared. The Astrophysical Journal. 1985;**298**:114-127

[17] Giommi P, Padovani P, Polenta G, et al. A simplified view of blazars: Clearing the fog around long-standing selection effects. Monthly Notices of the Royal Astronomical Society. 2012;**420**:2899-2913

[18] Ghisellini G, Maraschi L. Bulk acceleration in relativistic jets and the spectral properties of blazars. The Astrophysical Journal. 1989;**340**:181-189

[19] Fossati G, Maraschi L, Celotti A, et al. A unifying view of the spectral energy distributions of blazars. Monthly Notices of the Royal Astronomical Society. 1998;**299**:433-451

[20] Georgranopoulos M, Marscher AP. A viewing angle - kinetic luminosity unification scheme for BL Lacertae objects. The Astrophysical Journal. 1998;**506**:621-663

[21] Griffin MJ, Abergel A, Abreu A, et al. The Herschel-SPIRE instrument and its in-flight performance. Astronomy and Astrophysics. 2010;**518**:L3-L9

[22] Jenness T, Robson EI, Stevens JA. Observations of flat-spectrum radio sources at λ 850 μm from the James clerk Maxwell telescope II. April 2000 to June 2005. Monthly Notices of the Royal Astronomical Society. 2010;**401**:1240-1255

[23] Hardcastle MJ, Worrall DM, Birkinshaw M, et al. Unifying B2 radio galaxies with BL lac objects. Monthly Notices of the Royal Astronomical Society. 2003;**338**:176-189

[24] Trussoni E, Capetti A, Celotti A, et al. A multi-wavelength test of the FR I - BL lac unifying model. Astronomy and Astrophysics. 2003;**403**:889-901

[25] Cassaro P, Stanghellini C, Bondi M, et al. Extended radio emission in BL lac objects I: The images. Astronomy and Astrophysics Supplement Series. 1999;**139**:601-647

[26] Rector TA, Stocke JT. The properties of the radio-selected 1JY sample of bl lacertae objects. Astronomy Journal. 2001;**122**:565-592

[27] Landt H, Bignall HE. On the relationship between BL Lacertae objects and radio galaxies. Monthly Notices of the Royal Astronomical Society. 2008;**391**:967-987

[28] Seal Braun P, Baruch JEF. Optical photometry using Bradford ronotic telescope. Astrophysics and Space Science. 2009;**321**:177

[29] Seal Braun P. Dust in BL lac objects and Fanaroff-Riley radio galaxies: Infrared region. Astrophysics and Space Science. 2015;**360**:A62-A77

[30] Seal Braun P. 2017, study of BL lac objects in far infrared and submillimetre region. Astrophysics and Space Science. 2017;**362**:133-138

[31] Zirbel EL, Baum SA. On the FR I/FR II dichotomy in powerful radio sources, radio luminosities. The Astrophysical Journal. 1995;**448**:521-547

[32] Mao LS. 2MASS observation of BL lac objects II. New Astronomy. 2011;**16**:503-529

[33] Abrahamyan HV, Mickaelian AM, Knyazyan AV. The IRAS PSC/FSC combined catalogue. Astronomy and Computing. 2015;**10**:99-109

[34] Wisotzki L. Quasar spectra and the K correction. Astronomy and Astrophysics. 2000;**353**:861-866

[35] Archibald FN, Dunlop JS, Hughes DH, et al. A submillimetre survey of the star- formation history of radio galaxies. Monthly Notices of the Royal Astronomical Society. 2001;**323**:417-447

[36] Padovani P, Giommi P, Abraham P, et al. Filling the infrared gap: ISO

observations of 1 Jy BL Lacertae objects. Astronomy and Astrophysics. 2006;**456**:131-143

[37] Neugeauer,G., Mathews, K., Soifer, B.T., et al., Infrared photometry of the nebulosity around quasars, APJ., 1985; 298:275-280

[38] Chen PS, Fu HW, Gao YF. 2MASS observation of BL lac objects. New Astronomy. 2005;**11**:27-42

[39] Helou G, Khan IR, Malek L, et al. Iras observations of galaxies in the virgo cluster area. The Astrophysical Journal. 1988;**68**:151-172

[40] Dunne L, Eales S, Edmunds M, et al. The SCUBA local universe galaxy survey I. First measurements of the submillimetre luminosity and dust mass functions. Monthly Notices of the Royal Astronomical Society. 2000;**315**:115-138

[41] Siebenmorgen R, Freudling W, Krügel E, et al. ISOCAM survey and dust models of 3CR radio galaxies and quasars. Astronomy and Astrophysics. 2004;**421**:129-158

[42] Lu RS, Shen ZQ, Krichbaum TP, Iguchi S, Lee SS, Zensus JA. The parsec-scale jet of PKS 1749+096. Astronomy and Astrophysics. 2012;**544**:A89-A110

[43] Kino M, Niinuma K, et al. Morphological transition of the compact radio lobe in 3C 84 via the strong jet–cloud collision. Ap J Letters. 2021;**920**:L24-L30

[44] Dale DA, Aniano G, Engelbracht CW, et al. Herschel far-infrared and sub-millimeter photometry for the KINGFISH sample of nearby galaxies. The Astrophysical Journal. 2012;**745**:95-116

[45] Leeuw LL, Sansom AE, Robson E, et al. Observations of cold dust in nearby elliptical galaxies. The Astrophysical Journal. 2004;**612**:837-863

[46] Urry CM, Padovani P. Unified schemes for radio-loud active galactic nuclei. Publications of the Astronomical Society of the Pacific. 1995;**107**:803-891

[47] Willott CJ, Rawlings S, Archibald E, et al. A submillimetre difference between radio galaxies and radio quasars: Evidence for quasar-heated dust and synchronized submillimetre and radio source activity. Monthly Notices of the Royal Astronomical Society. 2002;**331**:435-446

[48] Capetti A, Celotti A. Testing the FR I/BL lac unifying model with HST observations. Monthly Notices of the Royal Astronomical Society. 1999;**304**:434-444

[49] Hardcastle MJ, Worrall DM. Radio, optical and X-ray nuclei in nearby 3CRR radio galaxies. Monthly Notices of the Royal Astronomical Society. 2000;**314**:359-364

[50] Shi Y, Ogle P, Rieke H, et al. Aromatic features in agn: Star-forming infrared luminosity function of agn host galaxie. The Astrophysical Journal. 2007;**669**:841-864

[51] Zhang, Bing-Kai, Zhang, Kai-Yin, YU, Li-Zhi et al., The long-term color variabilityof the BL lac object OQ 530, Astronomy Reports 2011;55:1074-1077

[52] Müller SAH, Haas M, Siebenmorgen R, et al. Dust in 3CR radio galaxies: On the FR 1 – FR 2 difference. Astronomy and Astrophysics. 2004;**426**:L29-L33

[53] Plotkin RM, Anderson SF, Brandt WN, et al. The lack of torus emission from BL lacertae objects: An infrared view of unification with

wise. Astrophysical Journal Letters. 2012;**745**:L27-L32

[54] Wang JM, Cheng C, Li YR. Clumpy accretion onto black holes. I. Clumpy-ADAF structure and radiation. The Astrophysical Journal. 2012;**748**:147-172

Chapter 7

Applications of Terahertz Technology: A Comprehensive Review of Advances and Future Prospects

Ali Shekari, Seyyed Salman Seyyed-Afghahi and Ali-Asghar Ebrahimi

Abstract

Advances in terahertz (THz) technology are accelerating, bringing unique applications across diverse fields like healthcare, telecommunications, security, and scientific research. Here's an overview of recent advances and some prominent applications of THz technology. Terahertz technology is still evolving, with applications constantly expanding as hardware becomes more efficient and cost-effective. As advances continue, THz technology is set to reshape industries, bridging the invisible frontier between electronics and photonics and offering unprecedented capabilities in data transmission, diagnostics, and material analysis. The integration of THz technology with other emerging fields, such as quantum computing and artificial intelligence, is also explored, offering insights into the transformative role THz will play in future technological landscapes.

Keywords: terahertz, optoelectronic, quantum, THz imaging, photonics

1. Introduction

Terahertz (THz) technology, encompassing electromagnetic waves, has emerged as a promising area of research due to its distinctive characteristics and potential applications across a range of fields. Positioned between the microwave and infrared regions of the electromagnetic spectrum, THz waves exhibit unique properties, such as non-ionizing energy, substantial penetration abilities through non-metallic materials, and high spectral resolution. These features make THz technology advantageous for applications where traditional imaging and sensing methods face limitations [1–3].

In recent years, advancements in terahertz generation and detection have paved the way for practical applications in medical diagnostics, security screening, quality control, high-speed communication, material characterization, and even astronomical observations. In medical imaging, for example, terahertz waves offer non-invasive and high-resolution imaging of tissues, providing a safe alternative for detecting surface abnormalities like skin cancer. Similarly, THz waves' capability to penetrate materials like plastics and clothing makes them ideal for security applications, such as detecting concealed weapons without harmful radiation exposure [4].

Despite these promising applications, the deployment of THz technology faces technical challenges. Efficient generation, detection, and signal transmission of THz waves require advanced, often costly, equipment and atmospheric absorption limit their effective range. Current research focuses on overcoming these challenges to unlock the full potential of THz technology.

This paper will explore the fundamental principles of terahertz technology, review its state-of-the-art applications, and address the ongoing challenges and future prospects in this rapidly evolving field.

2. Advances in terahertz technology

Recent advances in terahertz (THz) technology have broadened its applications and improved its practical utility across various sectors. The development of THz technologies has advanced significantly over the past few decades, fueled by their applications in fields like communication, imaging, sensing, and spectroscopy. THz radiation lies in the frequency range of 0.1 to 10 THz (wavelengths from 30 μm to 3 mm) and bridges the gap between microwaves and infrared light (**Figure 1**).

Key developments include:

2.1 THz sources and detectors

THz sources and detectors are essential components in the field of terahertz technology. These devices are critical for applications in imaging, spectroscopy, communication, and sensing. Here's an overview:

Quantum cascade lasers (QCLs): Recent improvements in quantum cascade laser technology allow for more efficient THz generation. QCLs can produce high-powered THz radiation, making them essential for imaging and spectroscopy [6].

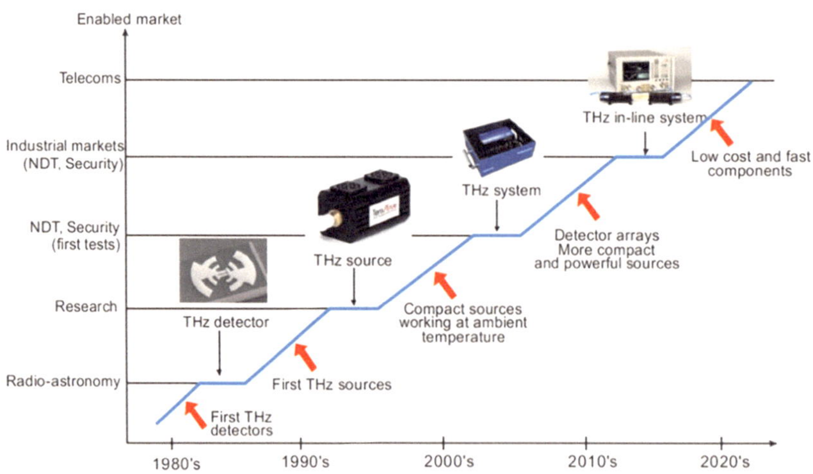

Figure 1.
Development of THz technologies [5].

Photoconductive antennas: These antennas use laser pulses to generate and detect THz waves and are increasingly sensitive, compact, and cost-effective.

Optoelectronic and semiconductor sources: Emerging semiconductors and materials such as graphene have improved THz wave generation, increasing the potential for on-chip THz technology [7].

THz detectors: Bolometers and field-effect transistor detectors have become more sensitive, providing precise measurements of THz radiation even in compact forms suitable for portable devices.

THz imaging systems: Driven by emerging environmental, societal, and economic challenges, the scientific and industrial sectors are striving to develop new materials or enhance existing ones across solid, liquid, and gaseous forms. Gaining a deep understanding of these materials and refining their manufacturing processes are crucial steps that rely on advanced technologies for measurement, inspection, and characterization at the core of material development.

Computational imaging: Advances in machine learning and computational imaging are being used to reconstruct THz images with higher clarity, even though complex or opaque materials [8].

2.2 THz communication systems

THz technology is poised to play a role in 6G networks, allowing ultra-fast data transmission due to the high-frequency band THz waves occupy. This potential extends beyond 5G's limits, particularly useful in dense urban areas or for high-capacity data transfer [9, 10].

Metamaterials for THz: The development of metamaterials that can manipulate THz waves has enabled better control and efficiency of THz wave propagation, with implications for both communications and sensing applications. Metamaterials offer transformative solutions for THz communication systems, addressing challenges like high attenuation, limited device efficiency, and beam control [11].

Their unique electromagnetic properties enable advanced functionalities, making them a cornerstone for developing next-generation THz communication technologies. Metamaterials reduce the size of antennas by achieving subwavelength focusing, allowing compact and lightweight solutions for THz transmitters and receivers.

Metamaterials can act as precise THz filters, enabling frequency division multiplexing (FDM) in communication systems, where multiple channels operate simultaneously without interference. By tailoring the electromagnetic response, metamaterial-based antennas improve directivity and minimize energy losses, essential for long-range communication.

Space communication: THz communications offer significant advantages for space applications due to their high bandwidth, small antenna size, and ability to support ultra-fast data rates. This emerging technology is poised to overcome the limitations of traditional radio frequency (RF) and optical communication systems in space.

High-speed data transfer between satellites in constellations, such as Earth observation or communication networks (e.g., Starlink, OneWeb). Supporting missions to Mars, the Moon, or beyond, where high data rates are needed to transmit detailed scientific data and images (**Figure 2**).

THz signals have short wavelengths, allowing the design of small and lightweight antennas, crucial for space missions where payload weight and size are limited. Unlike traditional RF bands, the THz spectrum is less congested, offering a relatively untapped frequency range for communication.

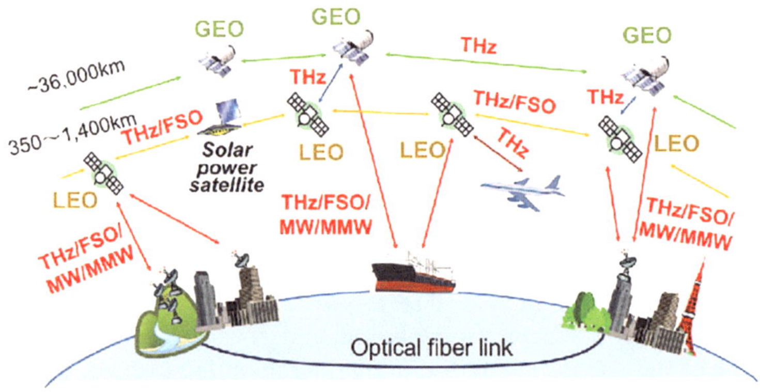

Figure 2.
THz communications for space communication [12].

THz beams are highly directional, improving link security and reducing interference, which is especially beneficial in crowded orbital environments. THz communications are harder to intercept or jam due to their narrow beamwidth and directional propagation.

For Earth-to-space links, atmospheric gases like water vapor significantly attenuate THz signals. This limitation makes THz communication more suitable for intersatellite or deep-space links where there is no atmospheric interference. Generating and amplifying THz signals efficiently is challenging due to the lack of mature electronic and photonic THz devices. THz communication could enable the next generation of space-based internet services, providing ultra-fast connectivity globally.

2.3 Medical diagnostics

THz technology is revolutionizing *medical diagnostics* by offering non-invasive, high-resolution imaging and spectroscopy capabilities. Unlike X-rays, THz radiation is non-ionizing, making it safe for biological tissues and repeated use. THz waves are highly sensitive to water and hydration levels, enabling the detection of subtle changes in tissues associated with diseases such as cancer and inflammation [13, 14]. THz spectroscopy can identify unique vibrational and rotational modes of biomolecules, aiding in the precise detection of chemical compositions and structures. Here's some application overview:

Cancer detection: THz imaging can identify early-stage skin and breast cancers due to its sensitivity to water and tissue properties, making it a valuable non-invasive tool for diagnostics.

Dental and skin health: THz imaging is used to detect decay in teeth and monitor burns or skin lesions, offering a radiation-free diagnostic alternative.

Biomolecular analysis: THz spectroscopy is applied to analyze biomolecular structures and interactions, aiding research in biochemistry and pharmacology (**Figure 3**).

2.4 Security screening and surveillance

Security screening and surveillance using THz technology is an advanced method for detecting concealed objects, substances, or threats. THz waves can penetrate common non-metallic materials like clothing, paper, and plastics, making it possible to detect hidden objects without physical inspection. THz waves are non-ionizing,

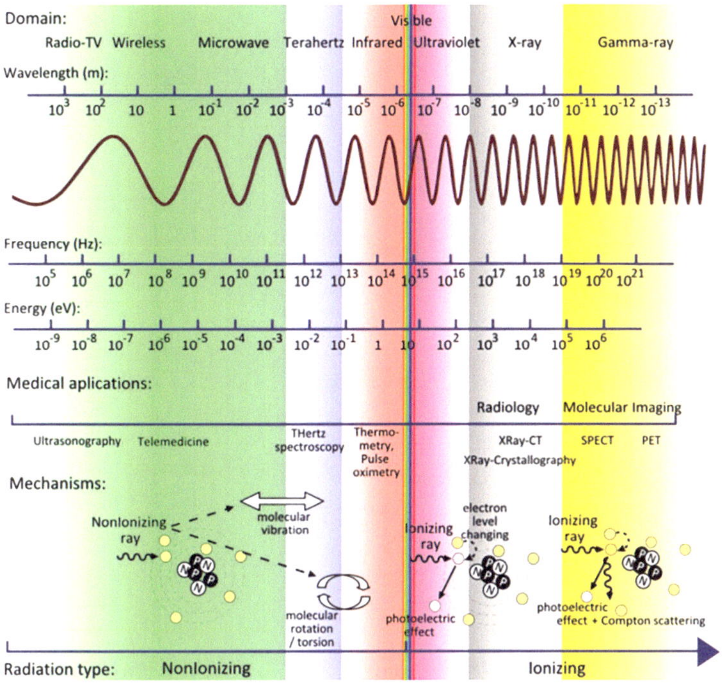

Figure 3.
The spectrum of electromagnetic waves, characteristics, and medical application [15].

making them safer compared to X-rays for scanning people and objects. THz technology can identify materials based on their spectral fingerprints, which is useful for detecting explosives, drugs, and other chemical compounds.

THz systems can produce high-resolution images, making them effective for identifying concealed objects. They can also be used in stand-off detection scenarios, allowing scans from a safe distance.

THz waves penetrate clothing and can detect concealed weapons or contraband in airports or at borders without exposing individuals to ionizing radiation. THz technology can identify chemical signatures of explosives or narcotics, providing a reliable, non-contact detection method for law enforcement and military use (**Figure 4**).

2.5 Material characterization and non-destructive testing (NDT)

Material characterization and non-destructive testing using THz technology is an emerging field that leverages the unique properties of THz waves to evaluate the properties, structure, and integrity of materials without causing damage [16].

THz waves can detect molecular vibrations and rotations, allowing for the identification of specific materials based on their spectral signatures. It is especially effective for distinguishing between similar-looking materials or compounds. Here are some applications in material characterization and NDT.

Quality control: THz imaging inspects material integrity in industries like aerospace, automotive, and electronics manufacturing, identifying defects in composite materials, coatings, or electronic components.

Pharmaceutical quality: THz spectroscopy analyzes drug formulations, allowing for precise quality control without damaging the products (**Figure 5**).

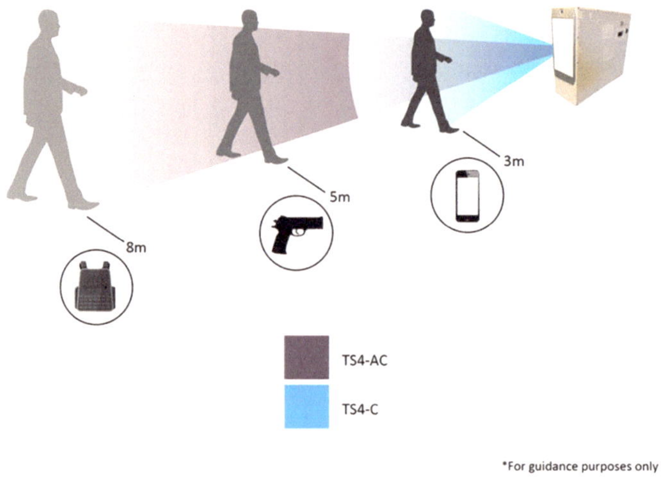

Figure 4.
Terahertz screening and tactical solutions.

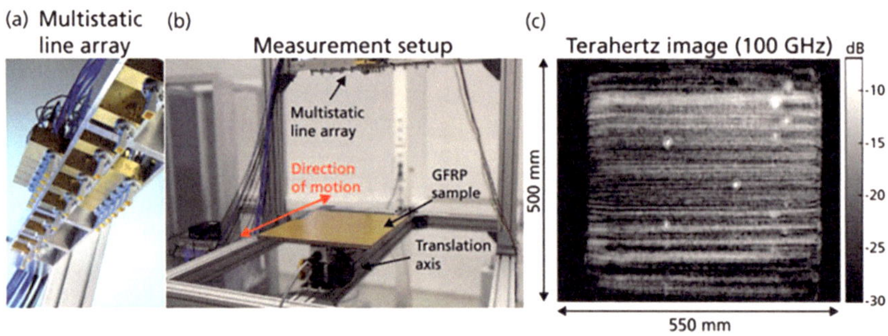

Figure 5.
Multistatic line array containing 12 transmitters and 12 heterodyne receivers. b Measurement setup with multistate array and translation axis for terahertz imaging. c Terahertz image of GFRP sample acquired with line array (feed motion of 19.2 cm/s) [17].

2.6 Telecommunications

Next-generation networks: The high bandwidth of THz frequencies promises data speeds far beyond 5G, potentially making real-time, high-resolution streaming and remote work more efficient.

Indoor high-speed networks: THz networks could enable faster data speeds for high-traffic locations like airports, stadiums, or conference centers.

2.7 Scientific research and astronomy

THz technology plays a significant role in *scientific research and astronomy* due to its ability to explore phenomena in the THz frequency range. This range provides unique insights into the physical and chemical properties of matter and the universe.

Many molecules exhibit unique absorption and emission features in the THz range, enabling precise identification of substances. THz spectroscopy reveals information about low-energy vibrational modes in crystals, polymers, and biomolecules.

At low temperatures, THz can probe quantum states and transitions in materials like superconductors and semiconductors. THz telescopes can observe cold interstellar dust, young stars, and other distant cosmic phenomena that are undetectable by other wavelengths. In laboratories, THz spectroscopy is used to investigate molecular structures and dynamics, assisting research in physics, chemistry, and materials science.

THz waves are used to observe molecular clouds and star-forming regions, as they penetrate dust and gas that block visible and infrared light. Hydrogen and carbon monoxide (CO) emissions in the THz range are key indicators of star and galaxy formation. THz spectroscopy identifies complex organic molecules, hinting at the precursors of life.

THz systems operate in the submillimeter band, capturing high-resolution images of celestial objects like galaxies, nebulae, and black holes. THz imaging excels in studying cold celestial bodies (10–300 K), including dust clouds and protostars.

On the other hand, ground-based observatories like the Atacama Large Millimeter/submillimeter Array (ALMA) employ THz technology. Space telescopes, such as the Herschel Space Observatory, are designed to operate in the THz range for unobstructed observations above Earth's atmosphere.

3. Challenges and limitations

Here we present the most important challenges and limitations in terahertz technology.

3.1 Efficient terahertz generation and detection

Limited source power: Many THz sources, such as photoconductive antennas and quantum cascade lasers, are limited in output power. This can restrict the range and resolution of THz systems, making it difficult to generate clear images or accurate measurements over larger distances.

Detection sensitivity: Current THz detectors struggle with achieving high sensitivity, particularly in room-temperature conditions. Many detectors require cooling to reach acceptable levels of sensitivity, which raises costs and complexity, especially for portable or real-time applications.

Narrow frequency range and tunability: Most THz sources are not widely tunable across the terahertz spectrum, limiting their applicability in applications that require specific frequency ranges, such as precise spectroscopic analysis in materials and biomedical research.

3.2 Atmospheric absorption and propagation loss

High absorption by water vapor: Terahertz waves are heavily absorbed by water vapor in the atmosphere, which causes substantial signal loss over long distances. This atmospheric attenuation limits the effectiveness of THz systems in outdoor applications, particularly in communication and environmental sensing.

Scattering in complex environments: In industrial settings or areas with high particulate matter, THz waves may scatter, which reduces signal clarity and reliability.

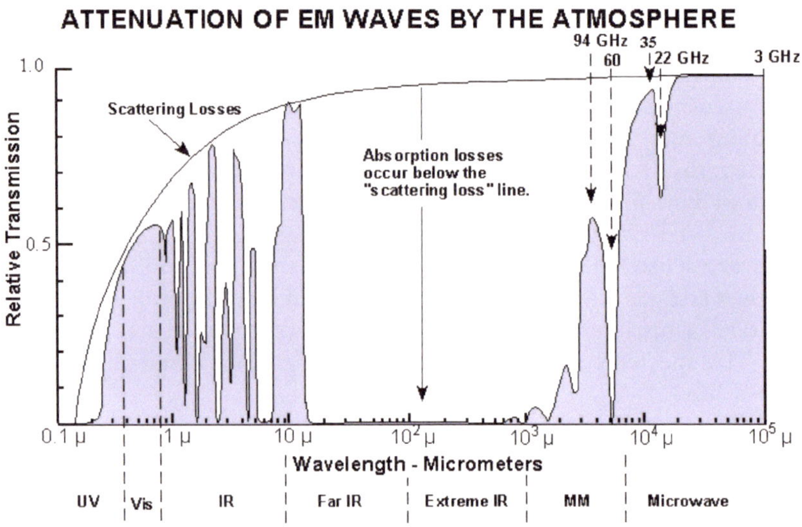

Figure 6.
Atmospheric absorption of electromagnetic radiation [18].

Addressing this scattering effect is essential for broader THz applications in real-world environments (**Figure 6**).

3.3 Resolution and penetration depth limitations

Limited penetration in certain materials: Although THz waves can penetrate materials like clothing, plastic, and paper, they face challenges in penetrating dense materials, metals, or highly conductive materials. This limits THz technology's utility in some security and industrial inspection applications [19].

Trade-off between resolution and depth: Higher-resolution THz imaging often sacrifices penetration depth. Achieving a balance between resolution and penetration depth is an ongoing challenge, particularly in fields requiring detailed imaging of internal structures, such as biomedical diagnostics or non-destructive testing (**Figure 7**).

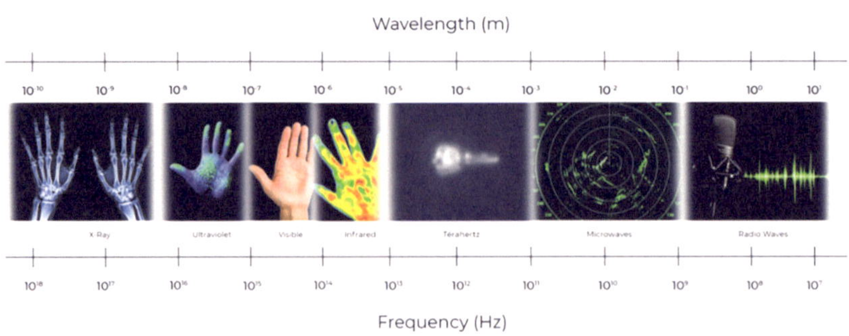

Figure 7.
Position of the terahertz region on the electromagnetic spectrum.

DOI: http://dx.doi.org/10.5772/intechopen.1008569

3.4 High equipment costs and complexity

Cost of THz components: THz systems remain expensive due to the costs of specialized materials, intricate assembly, and advanced manufacturing processes needed for THz sources and detectors. This high cost limits the accessibility of THz technology for research labs and industrial applications.

Complex system integration: Integrating THz components into compact, user-friendly systems that can function in real-time is challenging. Complex cooling requirements, calibration needs, and intricate system designs add to operational complexity and limit portability.

Scaling for commercial use: The lack of affordable, scalable solutions prevents THz technology from becoming mainstream. Efforts to simplify and mass-produce THz components could help make THz applications more commercially viable.

3.5 Data processing and interpretation challenges

Large data volumes: THz imaging and spectroscopy generate large datasets that require substantial processing power and storage. Managing and analyzing these data in real-time is challenging, especially for applications like airport security or industrial quality control.

Need for advanced algorithms: High-resolution THz data, particularly in spectroscopy, requires complex algorithms for accurate interpretation. Developing machine learning and AI models to analyze THz data effectively is essential, but it adds another layer of complexity to THz system deployment.

3.6 Regulatory and safety concerns

Lack of standardization: THz technology is still in its nascent stages, and there is a lack of standardized protocols for measurement, testing, and safety guidelines. The absence of clear regulatory standards hinders THz adoption, especially in sensitive fields like healthcare and security.

Safety and compliance issues: Although THz radiation is generally considered safe, there are concerns regarding prolonged exposure and high-power THz sources. Ensuring compliance with regulatory guidelines and addressing any potential health risks is essential for applications involving human interaction, such as medical diagnostics or public security.

3.7 Limited awareness and expertise

Need for cross-disciplinary expertise: THz technology requires knowledge from multiple fields, including optics, electronics, materials science, and data science. A limited pool of experts with this cross-disciplinary expertise restricts the speed of innovation and application development.

Lack of market awareness: Many industries are not yet familiar with the potential of THz technology, which slows its adoption. Increased awareness through targeted education and demonstration projects could help stimulate interest and demand.

3.8 Environmental and operational limitations

Sensitivity to environmental conditions: THz systems are sensitive to environmental factors, such as humidity and temperature fluctuations, which affect performance.

This limits THz technology's reliability in certain outdoor or uncontrolled environments.

Power consumption and battery life: High-power THz systems are energy-intensive, and achieving efficient, low-power THz devices remains a challenge, particularly for portable and remote applications.

4. Future directions

This section should highlight both the technical advancements required to overcome current limitations and the broader scope for new applications and integration with emerging technologies.

4.1 Advancements in THz generation and detection

Enhanced THz sources: Developing more efficient, compact, and cost-effective terahertz sources is crucial. Current generation techniques, such as quantum cascade lasers and photoconductive antennas, have made progress but require further optimization for broader accessibility, especially in portable applications [20].

Improved detectors: Sensitivity and resolution of THz detectors need enhancement to capture high-quality data in real time. Future research could focus on advanced materials (like graphene) and novel detection mechanisms that improve the signal-to-noise ratio and allow operation at room temperature, reducing cooling costs [21].

Frequency tunability: Development of widely tunable THz sources would enable specific applications, such as precise spectroscopic analysis and high-resolution imaging in medical and chemical research.

4.2 Miniaturization and portability

Portable and integrated THz systems: One promising direction is creating compact and integrated THz devices suitable for on-site testing and handheld applications. This would allow for broader use in security, medical diagnostics, and industrial quality control.

Microchip integration: Advances in integrating THz systems onto microchips (e.g., using CMOS-compatible technologies) could reduce both the size and cost of THz devices, paving the way for their incorporation into mobile and IoT devices.

4.3 Broadening applications in medical diagnostics

Non-invasive medical imaging: Research can focus on advancing THz imaging for deeper tissue penetration and higher resolution, potentially creating safer diagnostic tools that could complement or replace X-rays and MRIs for specific applications [22, 23].

Early disease detection: THz spectroscopy could be further developed for early diagnosis of diseases, identifying unique molecular signatures associated with cancer, skin diseases, or diabetes. With improved sensitivity, THz waves could detect subtle biochemical changes in tissues or cells at an early stage [24].

THz-integrated wearable devices: The possibility of integrating THz sensors into wearable health devices could enable continuous monitoring of skin hydration, biochemical composition, or wound healing, transforming personalized healthcare.

4.4 Expansion in wireless communication and 6G networks

THz communication networks: Terahertz waves are promising for next-generation 6G networks, offering extremely high data rates and low latency for data-intensive applications like real-time 3D rendering, virtual reality, and remote surgeries. Further development is needed in signal modulation techniques, transmission distance, and atmospheric interference reduction to make THz communication commercially viable.

Smart cities and IoT applications: Research can focus on using THz frequencies for high-density communication networks within smart cities and IoT ecosystems, enabling seamless data transmission across interconnected devices.

4.5 Integration with artificial intelligence and data analytics

AI-enhanced THz imaging and spectroscopy: Machine learning and AI can enhance the interpretation of complex THz data, improving accuracy in medical diagnostics, material characterization, and threat detection. AI-based image processing could help automate the analysis of THz images, identifying patterns and anomalies more precisely [25].

Big data analytics: The integration of big data analytics with THz systems can accelerate real-time decision-making, especially in large-scale applications like airport security or industrial quality control. Automated systems could leverage THz data for rapid sorting and detection of potential hazards or manufacturing defects.

4.6 Environmental and agricultural applications

Non-destructive agricultural monitoring: THz technology can be developed for assessing moisture content, pest detection, and structural quality of crops. This non-invasive monitoring could enhance yield prediction and quality control in agricultural industries [26].

Environmental monitoring: THz spectroscopy could expand into environmental applications, such as pollutant detection and monitoring atmospheric gases. Future research could explore the potential of THz sensors in identifying harmful compounds, offering a non-invasive and real-time alternative to traditional methods.

4.7 Space and astronomical applications

Advanced space-based THz observatories: Terahertz astronomy can benefit from space-based telescopes designed specifically for THz observation, bypassing atmospheric limitations that hinder ground-based telescopes. Future observatories could investigate interstellar gas clouds, star formation, and complex organic molecules in space with greater precision.

Planetary exploration: THz sensors on space missions could detect water ice, organic compounds, and other molecules on planetary surfaces or in atmospheres, expanding our understanding of celestial bodies and the conditions for life beyond Earth.

4.8 Material science and chemical research

Advanced THz spectroscopy for material analysis: With more precise tunable sources, THz spectroscopy can reveal detailed structural information about complex materials, like polymers, nanomaterials, and biomolecules. This could enable innovations in pharmaceuticals, chemical engineering, and nanotechnology [27].

Non-destructive testing (NDT): Enhanced THz imaging and spectroscopy for NDT could benefit industries such as aerospace and automotive manufacturing, allowing detailed internal inspection of components for defects without dismantling or damaging them.

5. Conclusion

The future of terahertz technology holds vast potential across numerous sectors, from healthcare and communications to space exploration and environmental science. While significant advancements in THz generation, detection, and portability are essential, further integration with AI and data analytics could unlock transformative applications. Ongoing research into cost reduction, improved sensitivity, and miniaturization will be key to making THz technology a mainstream tool for commercial and scientific purposes. Through interdisciplinary efforts, the terahertz domain is set to become an integral part of the next technological revolution.

Conflict of interest

The authors declare no conflict of interest.

Author details

Ali Shekari*, Seyyed Salman Seyyed-Afghahi and Ali-Asghar Ebrahimi
Faculty of Materials Science and Engineering, Department of Engineering, Imam Hossein University, Tehran, Iran

*Address all correspondence to: ali.shekari@yahoo.com

IntechOpen

© 2024 The Author(s). Licensee IntechOpen. This chapter is distributed under the terms of the Creative Commons Attribution License (http://creativecommons.org/licenses/by/4.0), which permits unrestricted use, distribution, and reproduction in any medium, provided the original work is properly cited. (cc) BY

References

[1] Van der Weide D. Applications and outlook for electronic terahertz technology. Optics and Photonics News. 2003;**14**(4):48-53

[2] Sharma V, Arya D, Jhildiyal M. Terahertz Technology and Its Applications. Electronics. 2021

[3] Hailu DM, Saeedkia D. Novel applications of terahertz technology for the plastic industry. In: 2016 41st International Conference on Infrared, Millimeter, and Terahertz Waves (IRMMW-THz). IEEE; 2016

[4] Hosako I et al. At the dawn of a new era in terahertz technology. Proceedings of the IEEE. 2007;**95**(8): 1611-1623

[5] Rogalski A. Progress in performance development of room temperature direct terahertz detectors. Journal of Infrared, Millimeter, and Terahertz Waves. 2022;**43**(9):709-727

[6] Kumar S. Recent progress in terahertz quantum cascade lasers. IEEE Journal of Selected Topics in Quantum Electronics. 2010;**17**(1):38-47

[7] Wen B, Ban D. High-temperature terahertz quantum cascade lasers. Progress in Quantum Electronics. 2021;**80**:100363

[8] Jansen C et al. Terahertz imaging: Applications and perspectives. Applied Optics. 2010;**49**(19):E48-E57

[9] Koch M. Terahertz communications: A 2020 vision. In: Terahertz Frequency Detection and Identification of Materials and Objects. NATO Science for Peace and Security Series B: Physics and Biophysics. Springer; 2007. pp. 325-338

[10] Federici J, Moeller L. Review of terahertz and subterahertz wireless communications. Journal of Applied Physics. 2010;**107**(11):111101-111101-22

[11] Xu W, Xie L, Ying Y. Mechanisms and applications of terahertz metamaterial sensing: A review. Nanoscale. 2017;**9**(37):13864-13878

[12] Nagatsuma T, Kasamatsu A. Terahertz communications for space applications. In: 2018 Asia-Pacific Microwave Conference (APMC). IEEE; 2018

[13] Amini T et al. A review of feasible applications of THz waves in medical diagnostics and treatments. Journal of Lasers in Medical Sciences. 2021;**12**

[14] Koch M. Bio-medical applications of THz imaging. In: Sensing with Terahertz Radiation. Part of the book series: Sensing with Terahertz Radiation. Springer; 2003. pp. 295-316

[15] Danciu M et al. Terahertz spectroscopy and imaging: A cutting-edge method for diagnosing digestive cancers. Materials. 2019;**12**(9):1519

[16] Wietzke S et al. Industrial applications of THz systems. In: International Symposium on Photoelectronic Detection and Imaging 2009: Terahertz and High Energy Radiation Detection Technologies and Applications. Proceedings of SPIE-The International Society for Optical Engineering; 2009:7385

[17] Ellrich F et al. Terahertz quality inspection for automotive and aviation industries. Journal of Infrared, Millimeter, and Terahertz Waves. 2020;**41**:470-489

[18] Ashfaq R. Study and Analysis of 5G Enabling Technologies, their Feasibility and the Development of the Internet of Things. Intelligence of Things: AI-IoT Based Critical-Applications and Innovations; 2021. pp. 101-143

[19] Fu X et al. Applications of terahertz spectroscopy in the detection and recognition of substances. Frontiers in Physics. 2022;**10**:869537

[20] Corsi C, Sizov F. THz and Security Applications: Detectors, Sources and Associated Electronics for THz Applications. Part of the book series: NATO Science for Peace and Security Series B: Physics and Biophysics (NAPSB). Springer; 2014

[21] Yachmenev A, Khabibullin R, Ponomarev D. Recent advances in THz detectors based on semiconductor structures with quantum confinement: A review. Journal of Physics D: Applied Physics. 2022;**55**(19):193001

[22] Movsisyan M et al. Applications of terahertz waves in medical diagnostics: A literature review. International Journal of Community Medicine and Public Health. 2024;**11**(6):2450

[23] Roh Y et al. Terahertz imaging with metamaterials for biological applications. Sensors and Actuators B: Chemical. 2022;**352**:130993

[24] Ajito K, Ueno Y. THz chemical imaging for biological applications. IEEE Transactions on Terahertz Science and Technology. 2011;**1**(1):293-300

[25] Settibathini V et al. Strategic analysis review of data analytics with the help of artificial intelligence. International Journal of Advances in Engineering Research. 2023;**26**:1-10

[26] Afsah-Hejri L et al. Terahertz spectroscopy and imaging: A review on agricultural applications. Computers and Electronics in Agriculture. 2020;**177**:105628

[27] Kojima S. Terahertz Spectroscopy in Advanced Materials Science. Trends in Terahertz Technology. Intechopen; 2023